AF712474

The Banfield Family Collection of Barometers

Also by the author and published by Baros Books:

Antique Barometers: an Illustrated Survey

Barometer Makers and Retailers 1660-1900

Barometers: Aneroid and Barographs

Barometers: Stick or Cistern Tube

Barometers: Wheel or Banjo

The Italian Influence on English Barometers from 1780

Also published by Baros Books:

Care and Restoration of Barometers by Philip R. Collins

A Treatise on Meteorological Instruments by Negretti & Zambra

The History of the Barometer by W.E. Knowles Middleton

The Banfield Family Collection of Barometers

Edwin Banfield

Baros Books

© Edwin Banfield 1995

First published 1995

Baros Books
5 Victoria Road
Trowbridge
Wiltshire
BA14 7LH

All rights reserved. No part of this publication may be reproduced, stored in a retrieval system, or transmitted, in any form or by any means, electronic, mechanical, photocopying, recording or otherwise, without the prior permission of the publisher.

British Library Cataloguing-in-publication Data
A catalogue record of this book is available from the British Library

ISBN 0 948382 10 4

Typesetting by Ex Libris Press, 1 The Shambles, Bradford on Avon, Wiltshire

Illustrations screened by Norton Photo Litho, New Pitt Cottages, Paulton

Printed and bound in Great Britain by Redwood Books, Trowbridge, Wiltshire

Contents

A NOTE ON THE COLLECTION

The Banfield Family Collection of Barometers began in a modest way in 1971 when I bought my first barometer in Fareham for £6: an onion-top rosewood instrument (illus. no. 176) in need of considerable repair. At the time I was a bank inspector in London and the south east and this allowed me to spend my lunch hours looking round the antique and junk shops near the branches I visited. A natural collector, with a longstanding interest in antique furniture, I had already filled our home with Victorian upholstered chairs and footstools in need of restoration, and my wife, Joan, was beginning to direct my attention to decorating the walls, which I initially did with antique maps. It was therefore perhaps inevitable that the onion-top barometer in Fareham should attract my attention.

Knowing nothing then about barometers, I took it to a local repairer, but he was too busy to undertake the work for some months. However, he explained what was involved in making the barometer work and I took it home and set about restoring it myself. The case was repaired, a new tube was fitted and the barometer worked; it was proudly hung on the wall and that was when my interest in barometers began.

During the next few months I bought three further wheel barometers, all different, all under £20 and all in need of repair. A little later, while working in Hartley Wintney, I noticed four wheel barometers hanging in a row. One had shell inlay which I had not seen on a barometer before, so I enquired the price. The antique dealer explained that the four were a dealer's job-lot and that he would only sell the four together for a total of £80. I bought them, doubling my number to eight and from then on started to accumulate a collection.

The limit of £20 was first exceeded when I came across a barometer in Weybridge by James Gatty of London (illus. no. 122), which looked rather special and was for sale at £32. I recalled that he was recorded as an important maker who had migrated from Italy to England around 1780, so I bought it. It was then that I started collecting stick barometers.

My wife remembers only too well the subsequent searches over the years for unusual barometers, particularly the hiring of dormobiles for touring holidays around Britain and the constant stops at antique shops, sometimes parked precariously on double yellow lines. Towns as far apart as Penzance, Broadstairs and Dundee were visited.

My original intention was to restore and then sell the barometers, but as I realised that every instrument was different in some way I could never bring myself to part with any. However, I did make one exception when I exchanged ten run-of-the-mill barometers for the John Halifax walnut barometer (illus. no. 119). The arrangement was agreed with a specialist barometer dealer who had bought a collection of eight important barometers and was being pressed by his bank manager to reduce his overdraft. I had already bought two of the collection and had no cash available to buy a third; the dealer was confident that he was likely to sell the ten more quickly than the expensive Halifax instrument.

By 1975 I had acquired a total of around a hundred stick, wheel and aneroid barometers, but I realised that there were still very many more different types available. I continued my search and over the years the collection increased to almost three hundred. I was excited to find a sympiesometer made by Alexander Adie (illus. no.109) who was the inventor, and in Tetbury I found a very rare pillar barometer by Brace Girdle (illus. no.173). The barometer by Thomas Rubergall (illus. no.52) has papers to show that it was owned by William Wordsworth and the balloonist's barometer by Henry Negretti (illus. no.100) is believed to have been used by his friend, James Glaisher the well-known balloon pioneer, for his early balloon experiments.

As the collection grew, so did the problem of where to house them and whether it would be possible to have them on public view. To this end, I began to look for suitable premises to exhibit the collection. This was partly realised in 1977 when the Churchill Gardens Museum in Hereford, where we were then living, agreed to take a hundred on loan. A more permanent solution presented itself when, in 1989, I met Philip Collins, barometer maker, restorer and enthusiast, of Merton in Devon. At that time, he was converting an old barn into workshops and a showroom and intimated that he would have spare accommodation available. The result, some five years later, was the opening of the Banfield Family Collection Museum in March 1995 by Peter Negretti, the great grandson of the famous Victorian barometer maker, Henry Negretti.

At the same time as I was acquiring barometers, I was amassing information about makers' names and dates and this led me on to further reasearch into the history and development of the barometer. In this way, I became an author as well as a collector. Space in this volume precludes anything but a very brief introduction to the various types of barometer in the collection. The interested reader, however, can find a good introduction to the subject in my *Antique Barometers: an Illustrated Survey*. A more comprehensive study can be found in the three companion volumes: *Barometers: Stick or Cistern Tube*, *Barometers: Wheel or Banjo* and *Barometers: Aneroid and Barographs*. To aid in the dating of barometers and the tracing of makers' names, I have also written *Barometer Makers and Retailers, 1600-1900*, an alphabetical list of more than 4,000 makers and retailers in England, Scotland, Wales and Ireland, including working dates, addresses where known and details of interest. My most recent book, *The Italian Influence on English Barometers from 1780* celebrates the importance of the work of the Italian migrant barometer makers in England from the late eighteenth century onwards, including the renowned partnership of Negretti and Zambra.

For those interested in repairing and restoring barometers I recommend *Care and Restoration of Barometers* by Philip Collins, the Museum Curator. He describes the art of restoration and gives advice on how the instruments should be handled and maintained.

A NOTE ON THE BAROMETER

The barometer was invented in 1643 by an Italian named Torricelli. He took a glass tube three feet long, sealed at one end, and filled it with mercury. He then placed his finger over the open end and inverted the tube into a container of mercury. On removing his finger, he found that the tube remained full of mercury to a height of around 29 inches above the surface of the mercury in the container. Further experiments soon proved that the weight of the column of mercury was proportionate to the pressure of air outside the tube and Torricelli discovered that the height of mercury was dependent on the pressure of the atmosphere.

Robert Boyle appears to have been the first person to introduce the barometer to England. He returned to England from Italy in 1644 with particulars of the experiment and used a graduated scale to record the height of mercury in the tube. From that date the barometer was developed as a scientific instrument, mainly for the measurement of heights, and it was not until about 1675 that barometers were made for domestic use; these were known as *stick or cistern tube barometers.*

It was soon realised that the variation in atmospheric pressure only varied the height of a column of mercury by a maximum of 3 inches, so scientists considered ways of extending the scale. Sir Samuel Morland in 1670 thought of the simple method of bending the tube almost to a right angle, just below the lowest point that the mercury could fall, so that for every inch the barometer would rise in a vertical position the mercury would extend several inches along an almost horizontal tube. These were called *angle or signpost barometers.*

Marine barometers were developed for use on board ship. The main problem was to prevent the oscillation of the mercury in the tube caused by the motion of the vessel. This was overcome by Edward Nairne in 1773 who constricted the tube to a very fine bore from the top of the cistern to just below the scale. This prevented the mercury from rising and falling rapidly as the ship rolled. To keep the barometer upright it was mounted on a gimbals. Barometers with cistern tubes were also developed with a greater accuracy to satisfy the demands of scientists who suggested various designs. The *Fortin* instrument was fitted with a glass cistern so that the level of the mercury could be closely observed, while *mountain barometers* were calibrated down to 20 inches to record heights up to 10,000 feet. *Farmers' and miners' barometers* also had special features, while the *Admiral Fitzroy barometers* record his remarks on weather interpretation.

Double-tube barometers using mercury and an oil were developed to extend the scale, and multiple-tube instruments were made in an endeavour to improve their portability by reducing their size; these comprised several parallel tubes joined together and filled alternately with mercury and a lighter liquid. The *sympiesometer*, patented by Alexander Adie in 1818, was another attempt at portability by reducing the barometer size. It consisted of a bulb cistern tube filled with hydrogen and a coloured almond oil which was also affected by air pressure.

The *wheel or siphon tube barometer* evolved from a desire to extend the scale so that a more accurate reading could be taken. The idea is attributed to Robert Hooke in 1663; his idea was to bend the open end of the tube into a U-shape and use the mercury level in the short limb to record air pressure changes. A glass float rests on the mercury and is connected by a silk cord over a pulley to a counterpoise glass float; the arbor of the pulley wheel is connected to a pointer over a dial and when the mercury moves up or down in the tube it will raise or lower the float which, in turn, will rotate the pulley. Italian immigrants to Britain towards the end of the eighteenth century popularised this type of barometer which was known as the wheel or banjo barometer.

In 1849 a French engineer, Eugene Bourdon, registered a patent for a *metallic barometer*. It consisted of a flattened tube of metal, exhausted completely of air, and curved round to make almost a circle, having an elliptical cross-section. With a change in air pressure the tube curled or uncurled slightly; one end of the tube was fixed to the frame, with the other end free, so that changes in the distance between the two ends could be magnified by a lever system connected to an indicating hand.

A much more popular metal or *aneroid barometer* was invented in 1843 by Lucien Vidie, another Frenchman. It consisted of a shallow, sealed metallic chamber, almost exhausted of air, with its upper and lower surfaces corrugated in concentric circles to make them flexible. The lower surface was fixed and the sides of the chamber were prevented from collapsing inwards by internal springs. The upper surface was in contact with the short arm of a lever and a longer arm operated an index pointer. Changes in air pressure caused changes in the height of the chamber and so moved the index pointer which was magnified by the lever system. The *barograph* is an aneroid barometer which also records, by a pen mechanism, a continuous graph of the air pressure so that trends can be easily seen.

Most barometers are attractive pieces of furniture and collecting them is a practical proposition as most are neither large nor heavy and require little or no attention. Unlike other furniture, they take up no floor space and a further attraction is that almost all are engraved with the maker's or retailer's name and often his address so that it is possible, with a little research, to discover the barometer's approximate age and, sometimes, interesting information about the makers who were, originally, mathematical, optical, or philosophical instrument makers or clockmakers.

As such a large variety of barometers were produced, it is unlikely that a collector will find any two specimens to be exactly the same. They have also proved to be a profitable investment as their values have increased over the years, some well in excess of the rate of inflation.

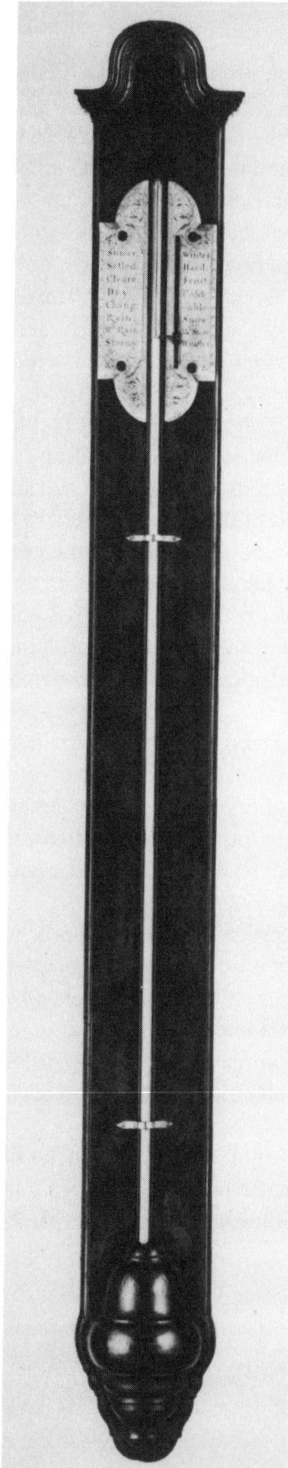

1

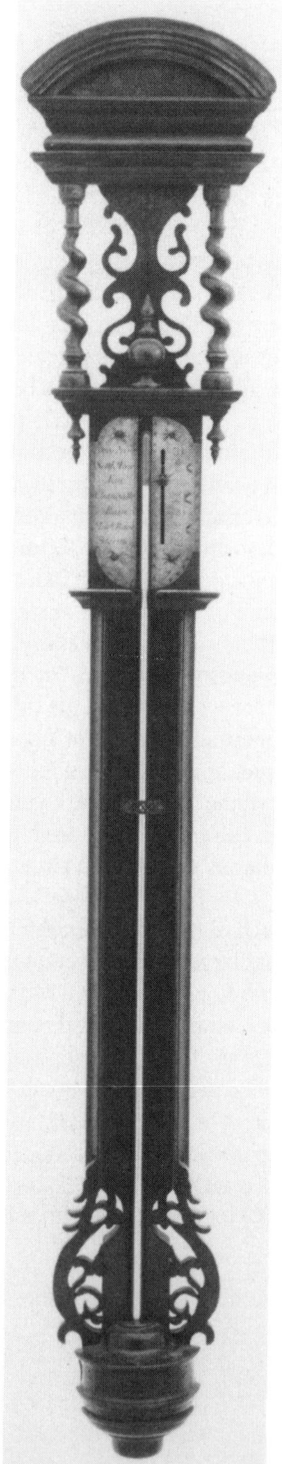

2

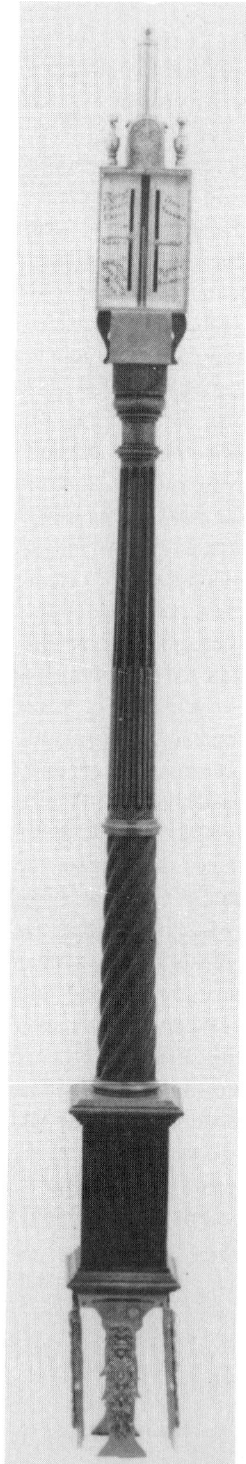

3

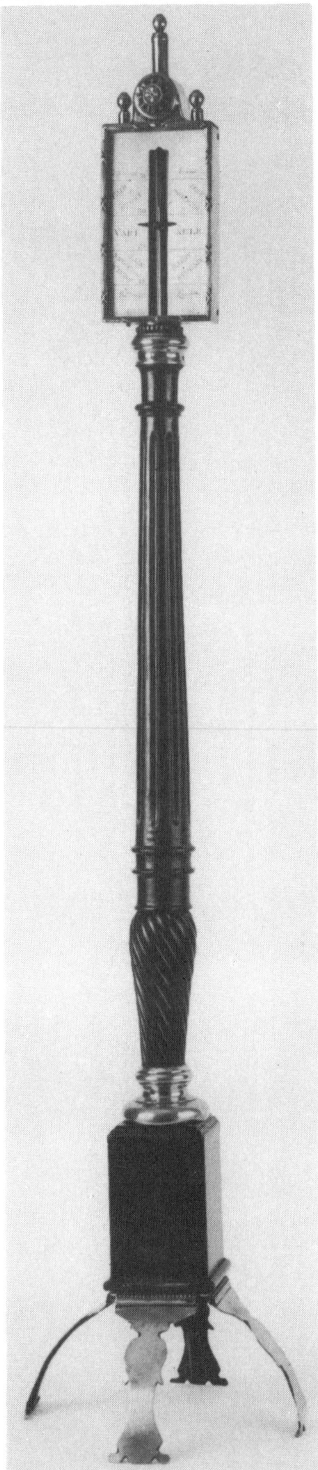

4

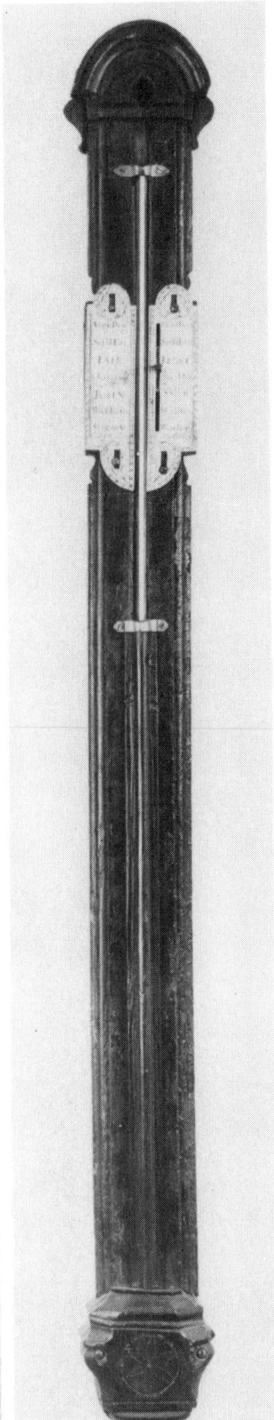

5

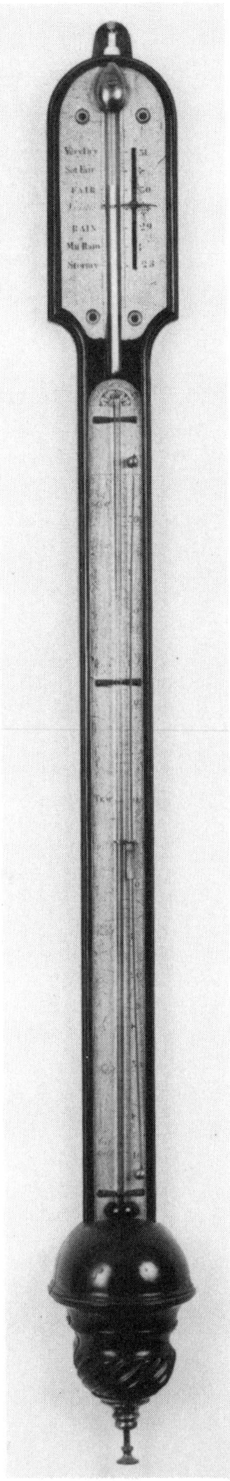

6

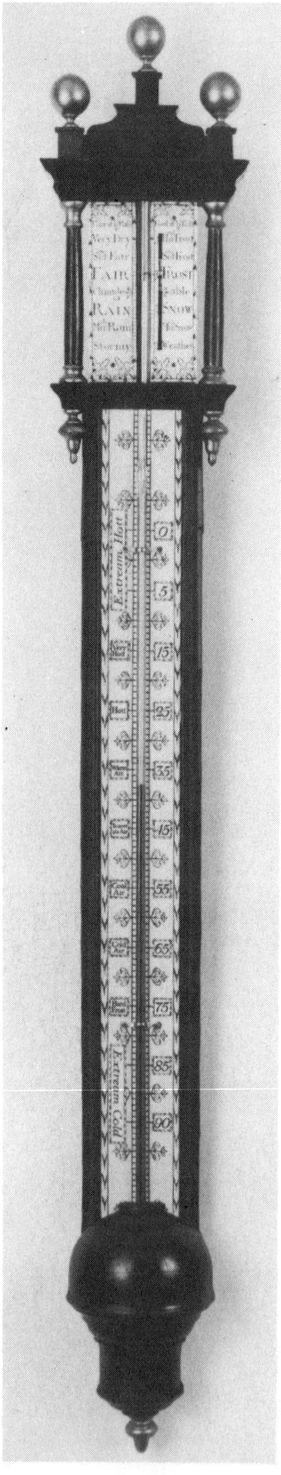
Very Dry
FAIR
RAIN
Stormy
FROST
SNOW
Weather
Extream Hott
Extream Cold
0
5
15
25
35
45
55
65
75
85
90

7

Fair if ☿ Rise
Very Dry
Set Fair
FAIR
Change
RAIN
Much Rain
Stormy
Foul if ☿ Fall
Hard Frost
Set Frost
FROST
able
SNOW
Much Snow
Weather
Extream Hott
0
5

8

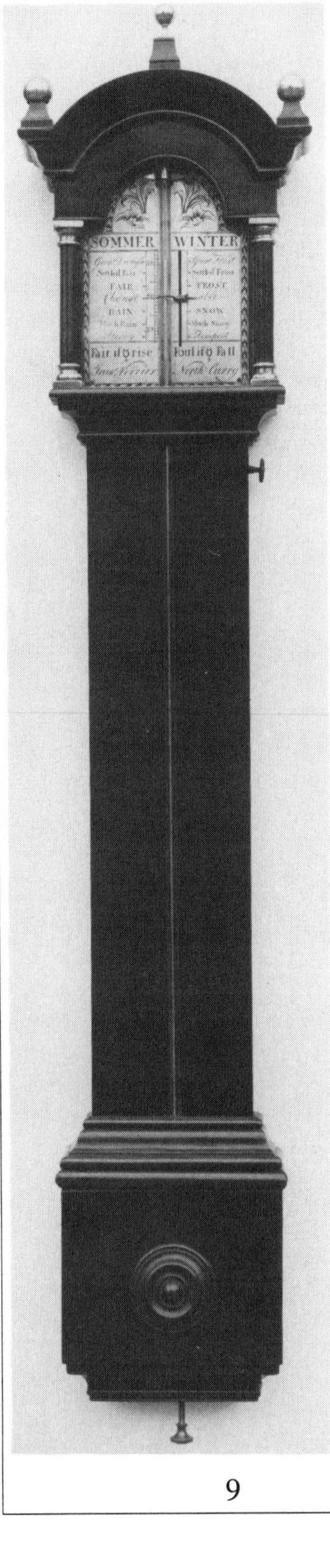
SOMMER
WINTER
Fair if ☿ rise
Foul if ☿ Fall
North Curry

9

SOMMER
WINTER
Great Drought
Settled Fair
FAIR
Change
RAIN
Much Rain
Stormy
30
29
28
Great Frost
Settled Frost
FROST
able
SNOW
Much Snow
Tempest
Fair if ☿ rise
Foul if ☿ Fall
Jam.s Verrier
North Curry

10

11

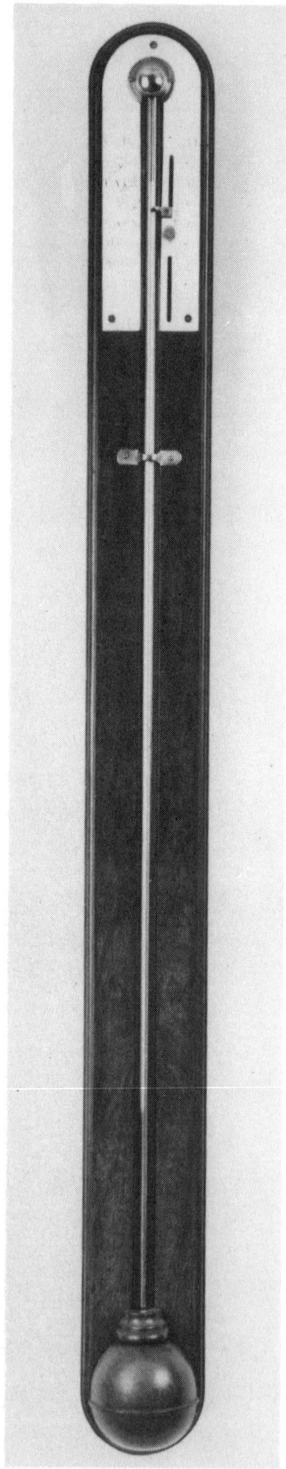

12

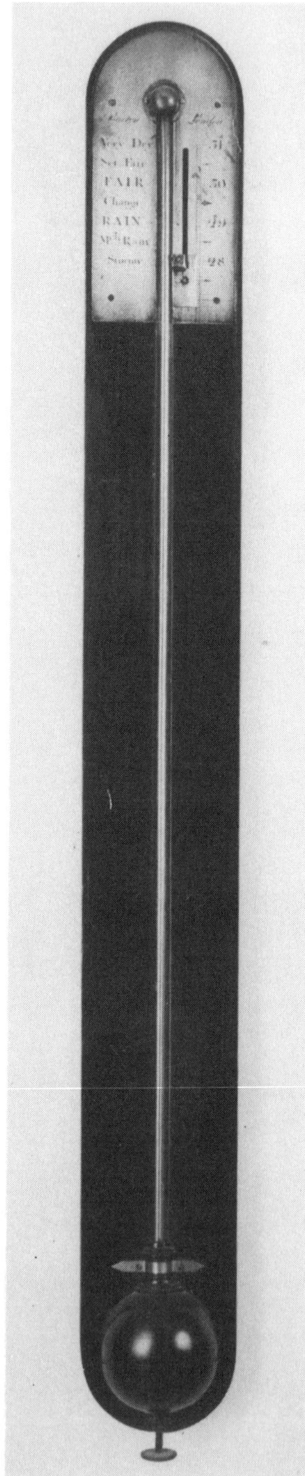

13

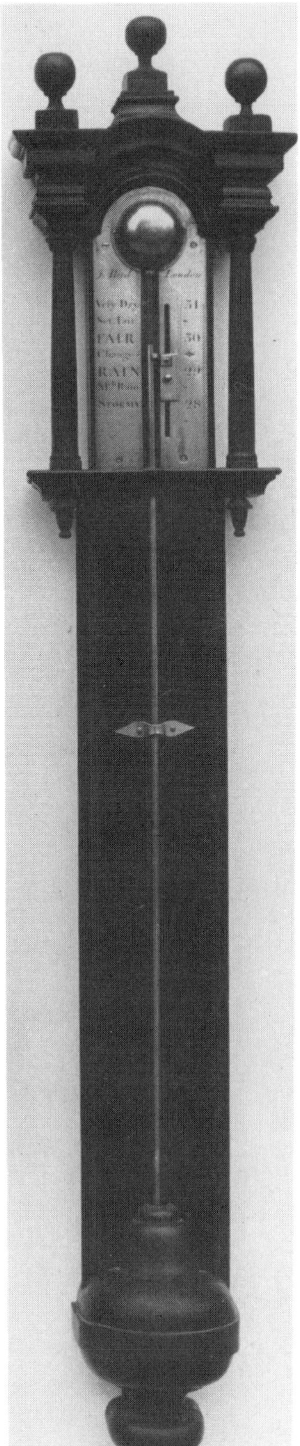

14

15

30
Frost
29
Snow
28
J.Rittson
Fecit

16

The
Great
BARO
METER
Drought
30
FAIR
Frost
Changeable
29
RAIN
Snow
28
Storm
J.Rittson
Fecit

17

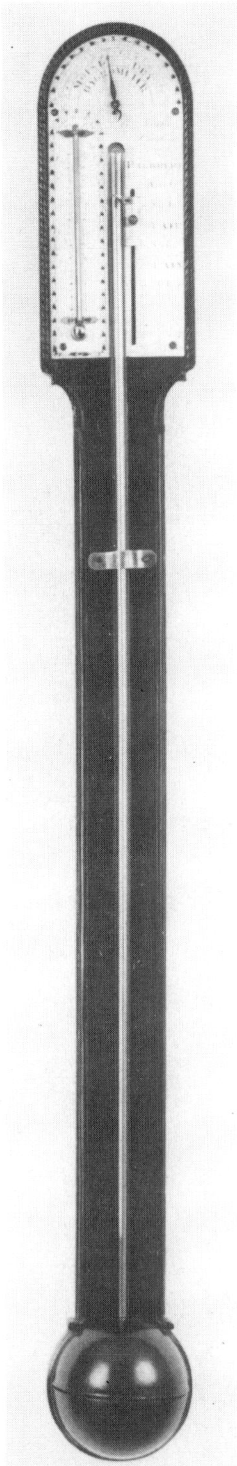

18

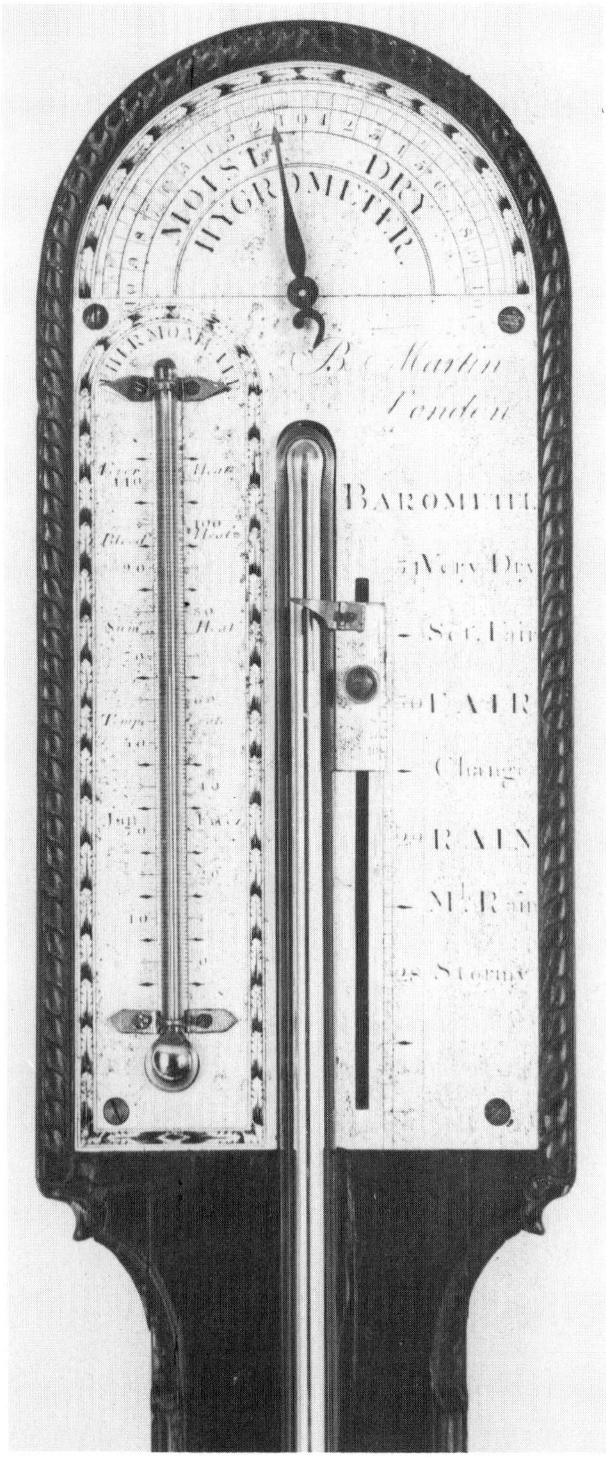
MOIST
DRY
HYGROMETER
B. Martin
London
Very Dry
Set Fair
FAIR
Change
RAIN
Stormy

19

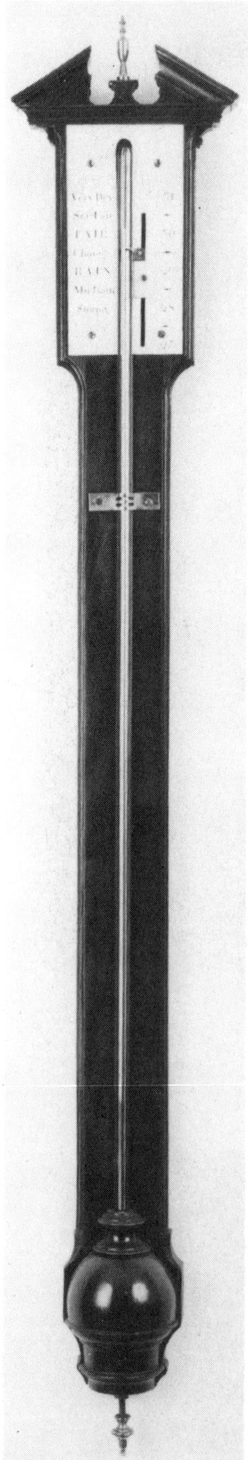

20

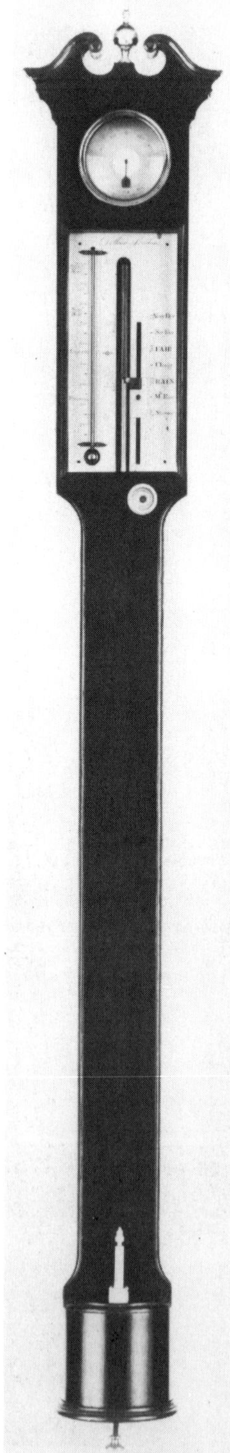

21

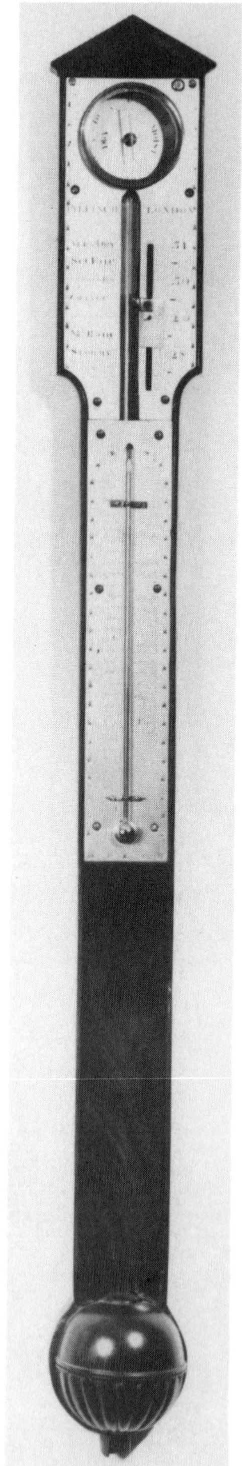

22

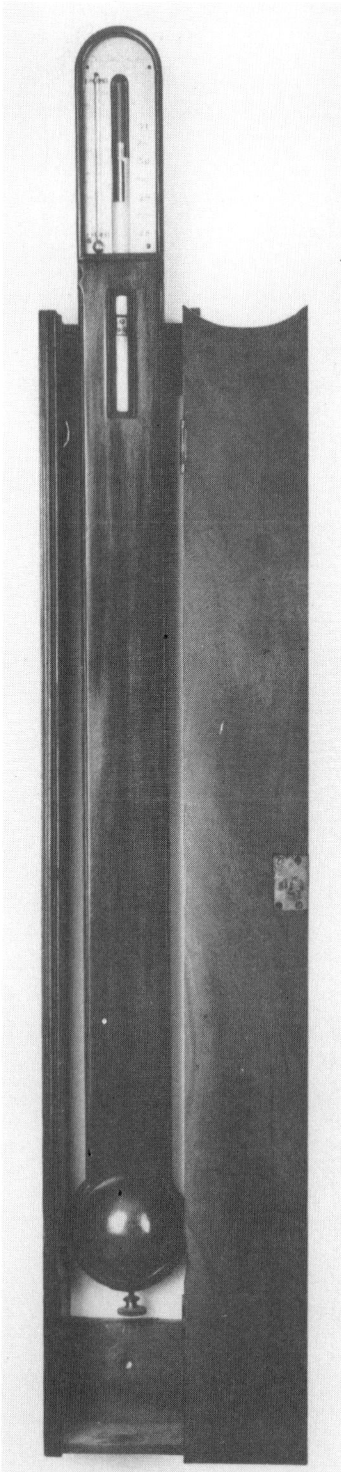

23

Troughton London
110
100
90
80
70
60
50
40
20
10
0
31
Fair
30
Change
29
Rain
28

24

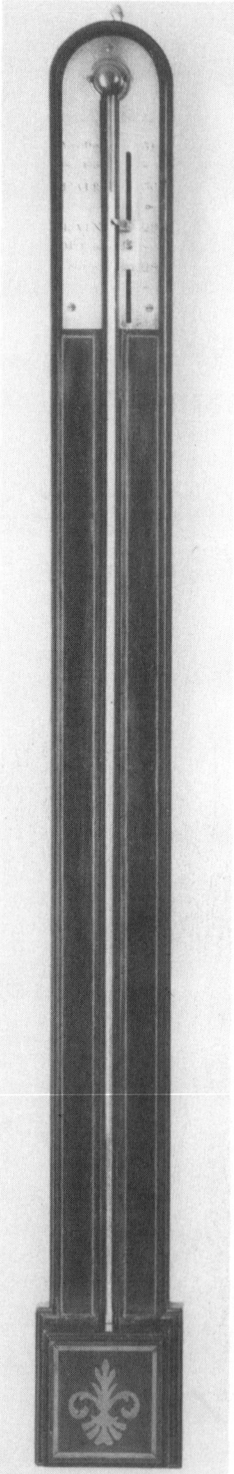

25

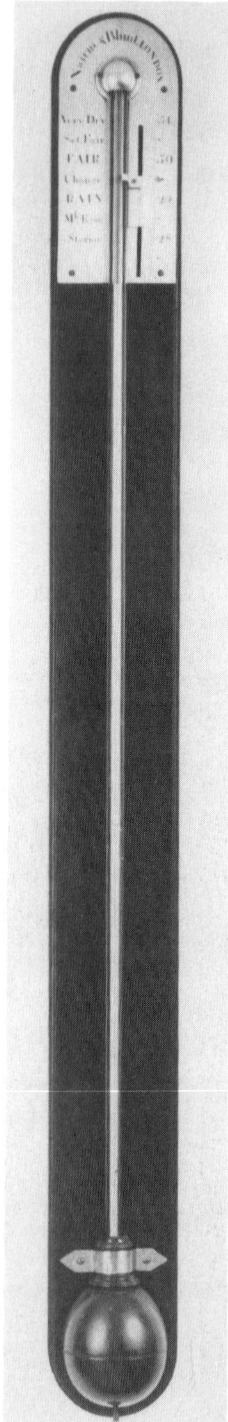

26

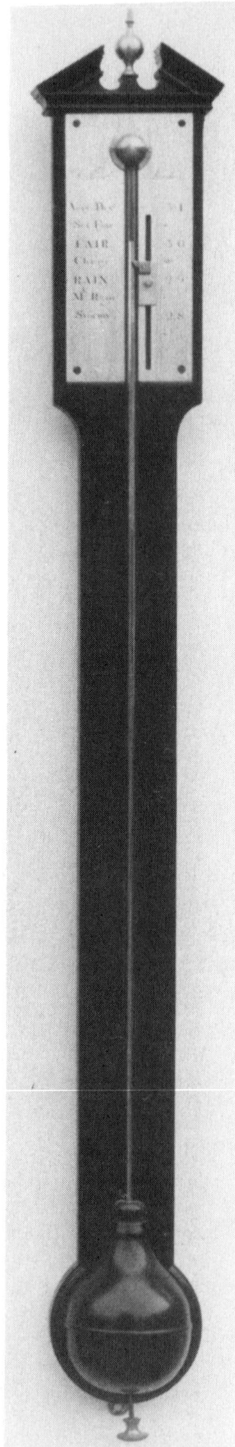

27

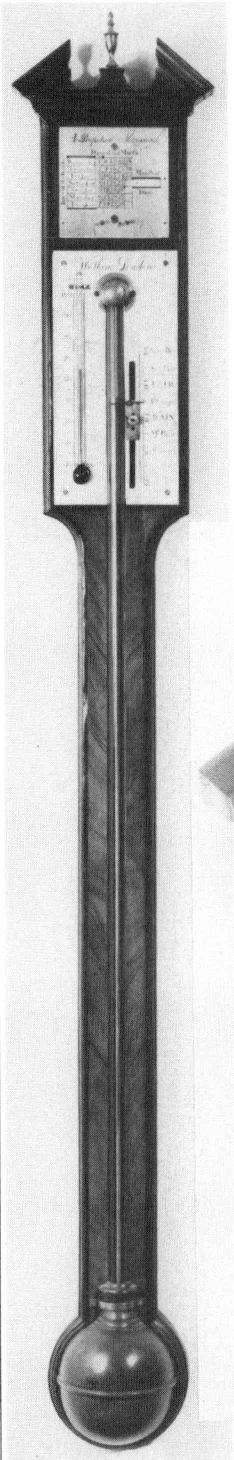

28

A Perpetual Almanack
Days of the Month
Days of the Week
Wednesday 1 8 15 22 29
Thursday 2 9 16 23 30
Friday 3 10 17 24 31
Saturday 4 11 18 25
Sunday 5 12 19 26
Monday 6 13 20 27
Tuesday 7 14 21 28
Month of
Sept. 30
Days
Watkins London

29

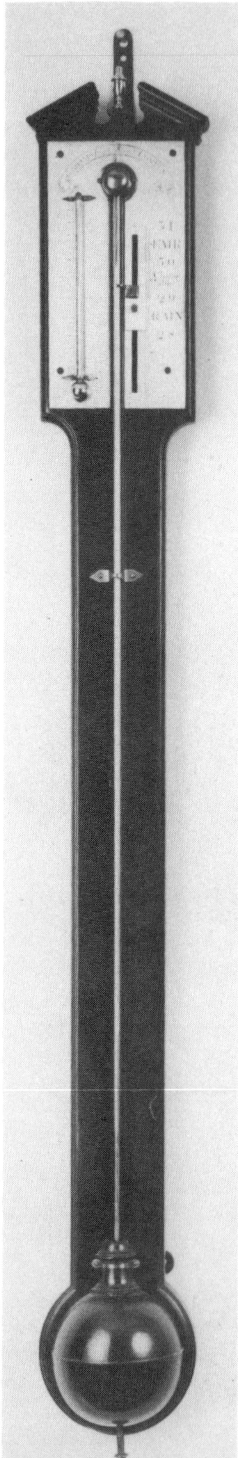

30

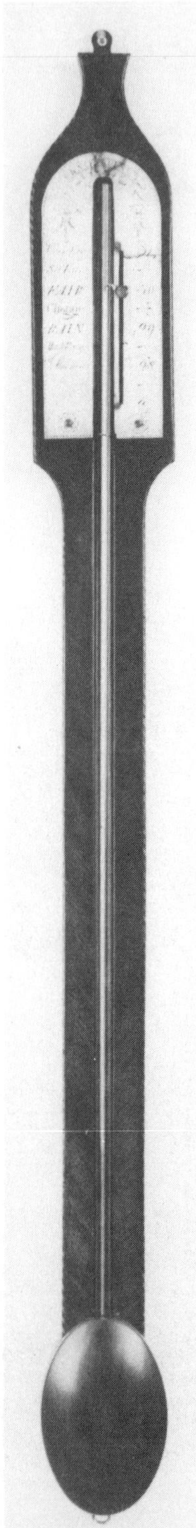

31

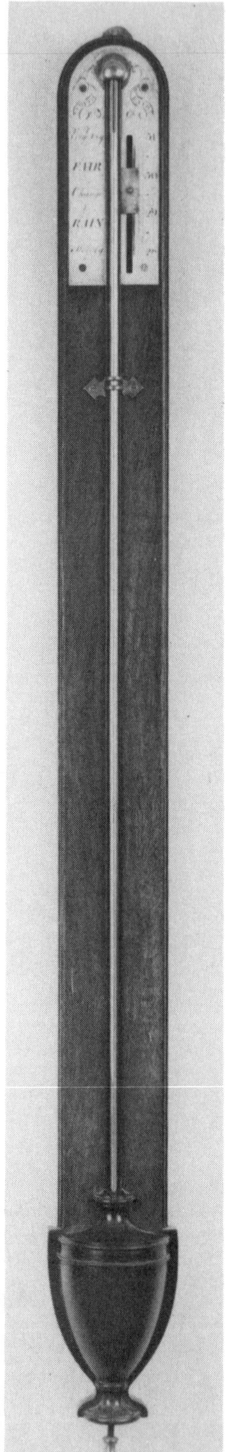

32

33

34

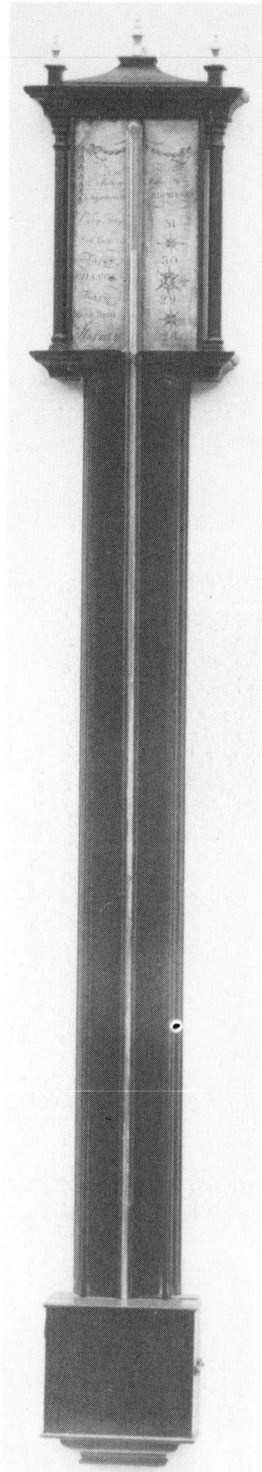

35

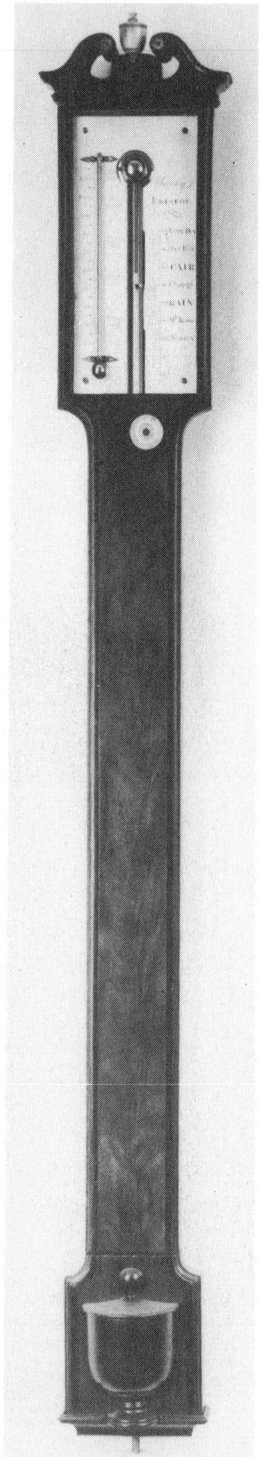

36

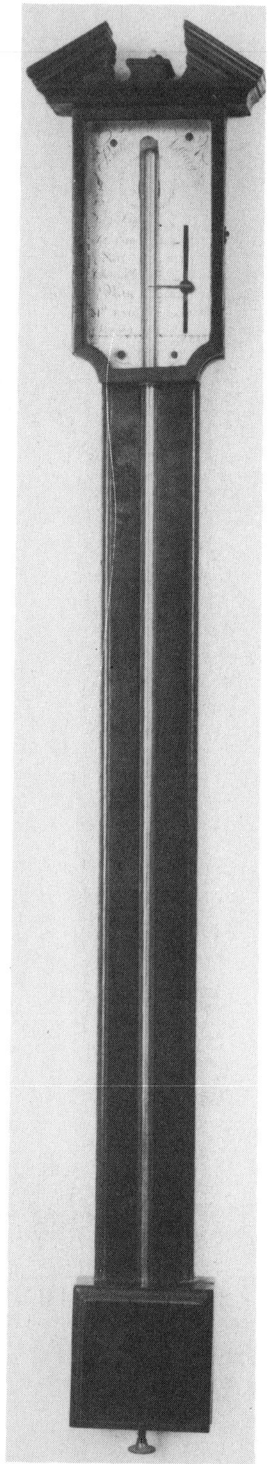

37

38

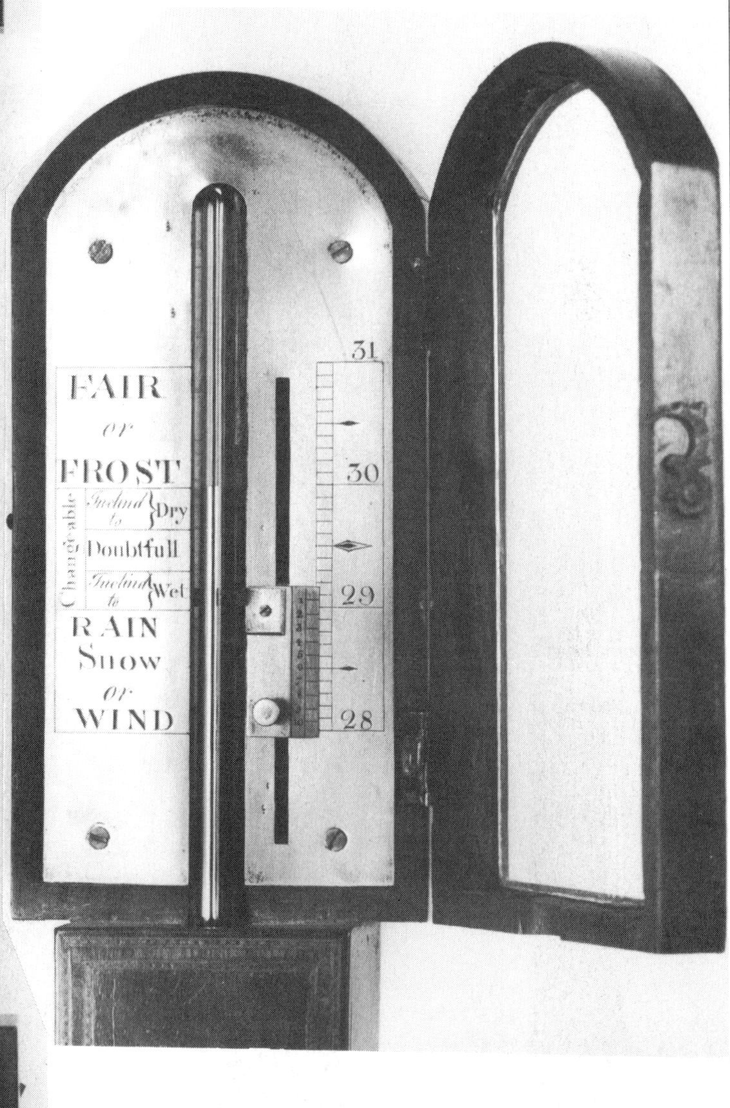
FAIR
or
FROST
Changeable
Inclind to
Dry
Doubtfull
Inclind to
Wet
RAIN
Snow
or
WIND
31
30
29
28

39

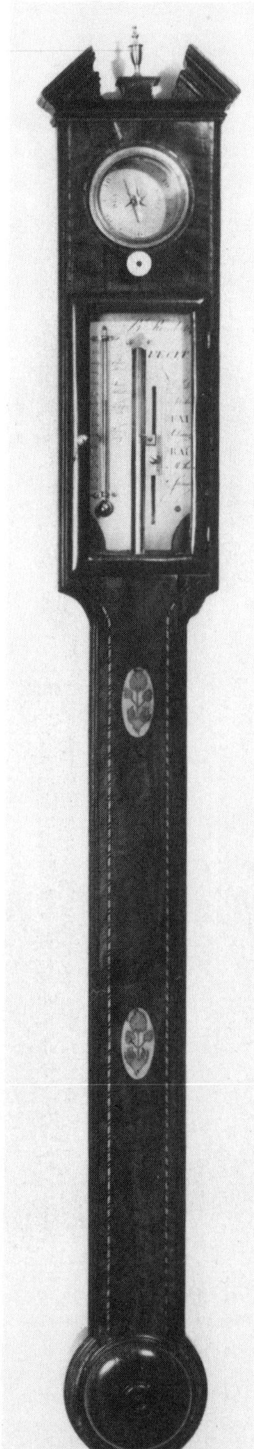
40

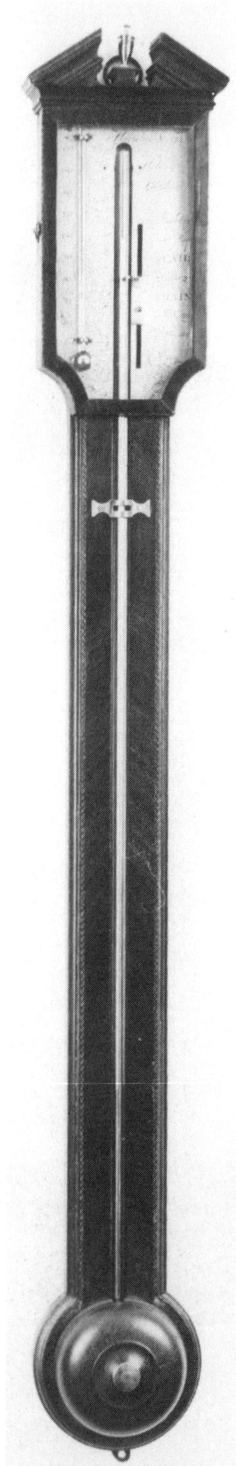
41

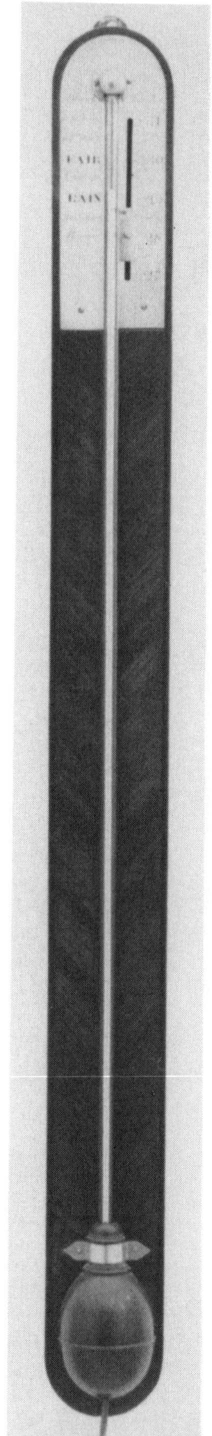
42

43

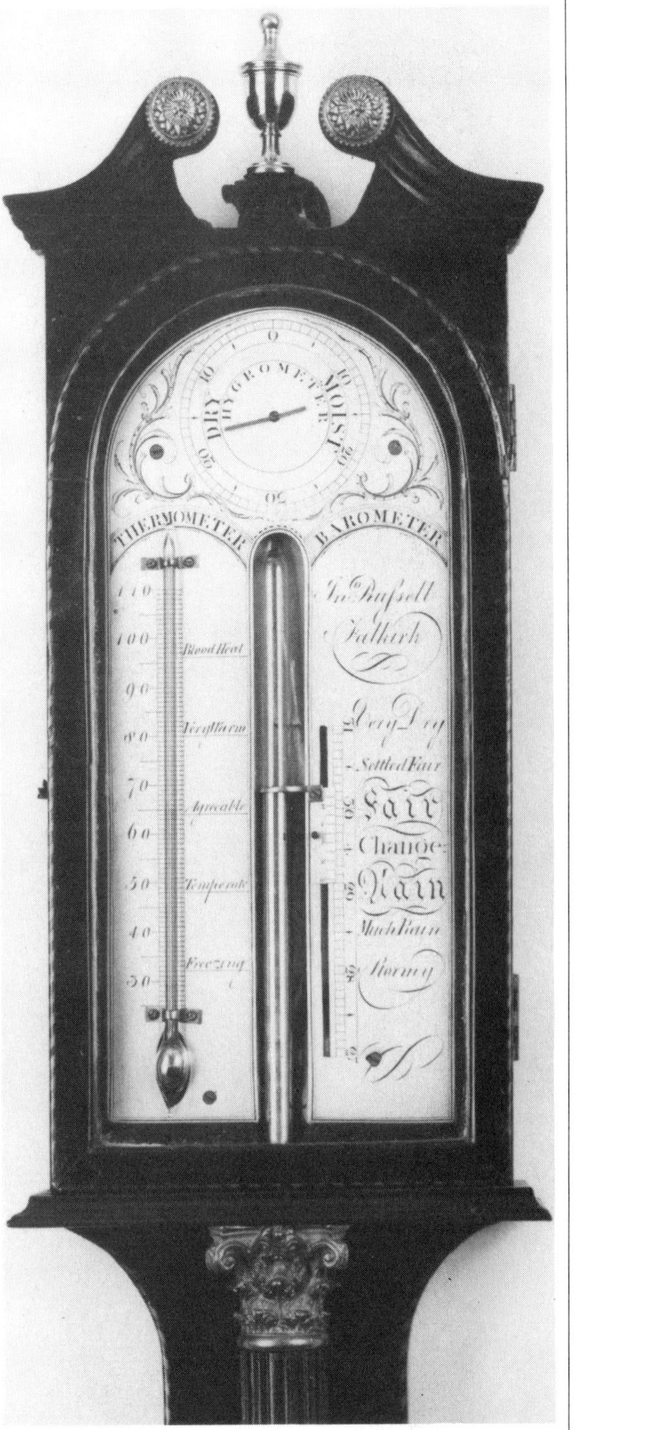
HYGROMETER
DRY
MOIST
THERMOMETER
BAROMETER
Jn.o Rufsell
Falkirk
Blood Heat
Very Warm
Agreeable
Temperate
Freezing
Very Dry
Settled Fair
Fair
Change
Rain
Much Rain
Stormy

44

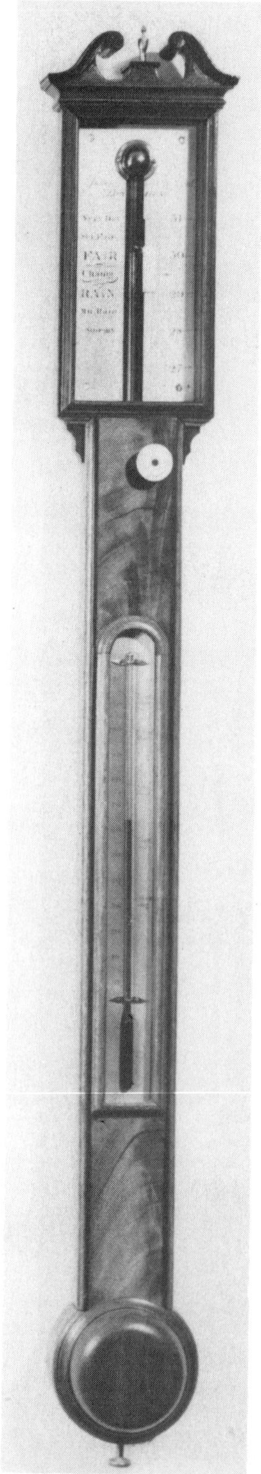

45

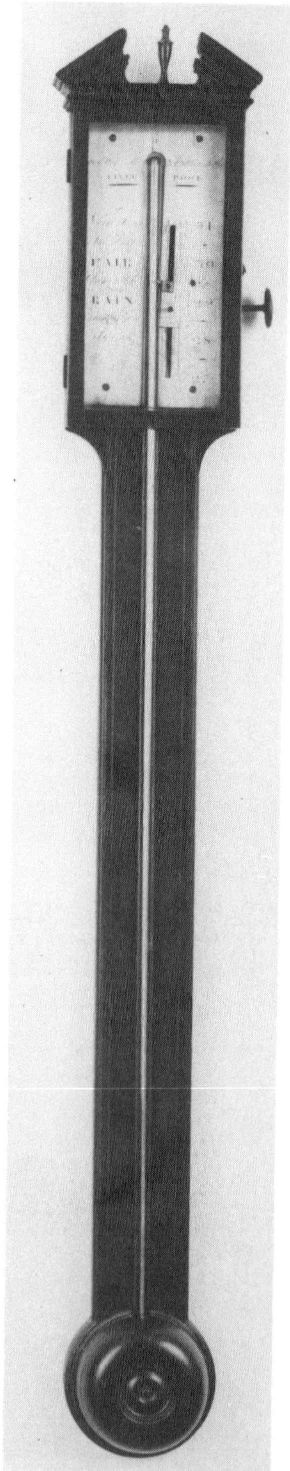

46

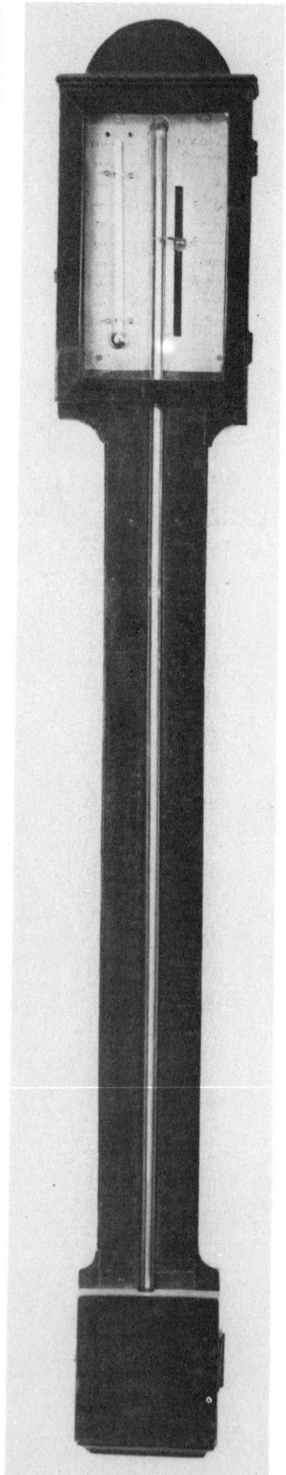

47

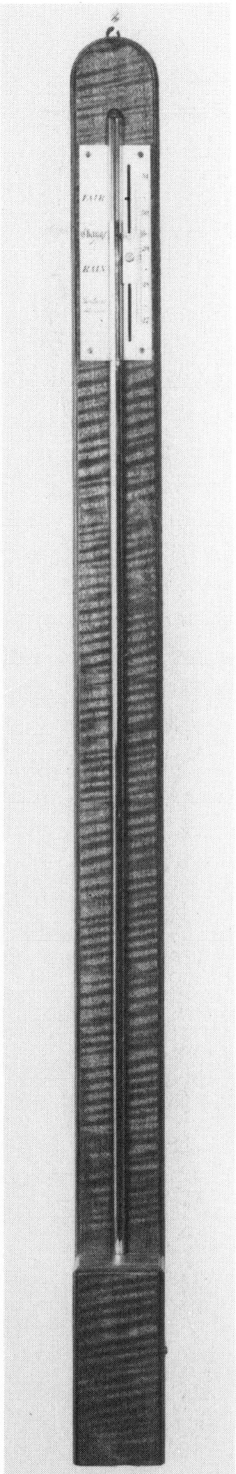

48

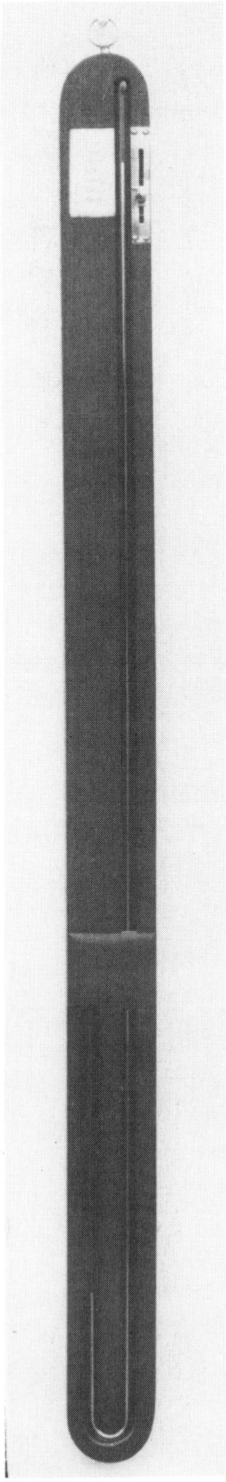

49

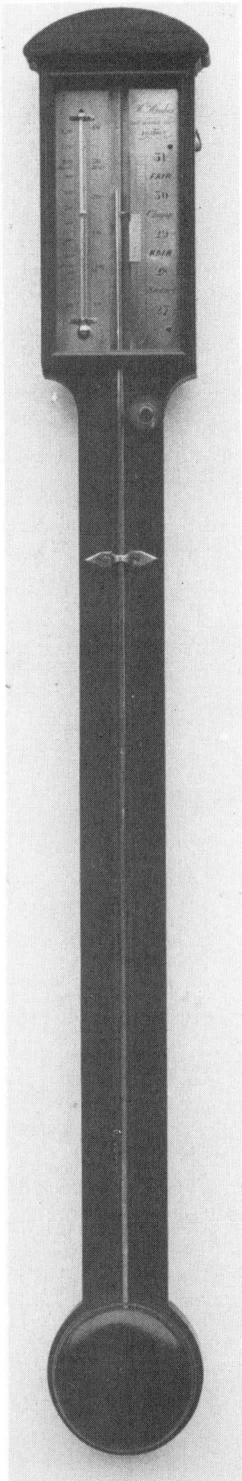

50

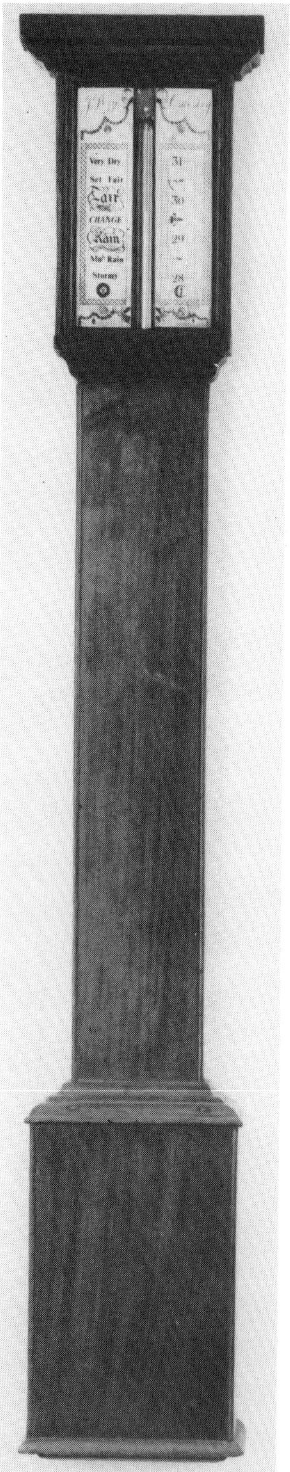
Very Dry
Set Fair
Fair
CHANGE
Rain
Mu^t Rain
Stormy
31
30
29
28

51

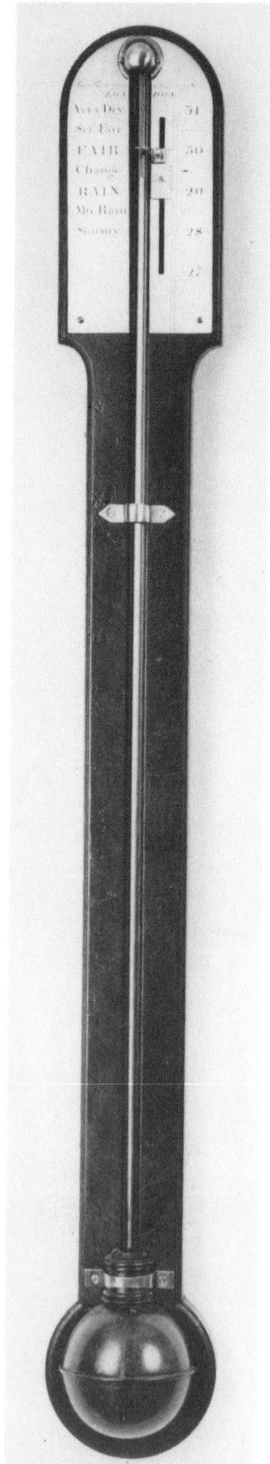
Very Dry
Set Fair
FAIR
Change
RAIN
Mu Rain
Stormy
31
30
29
28
27

52

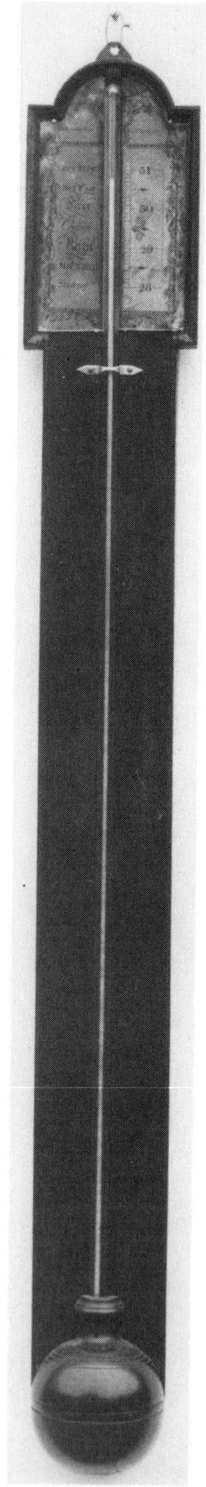

53

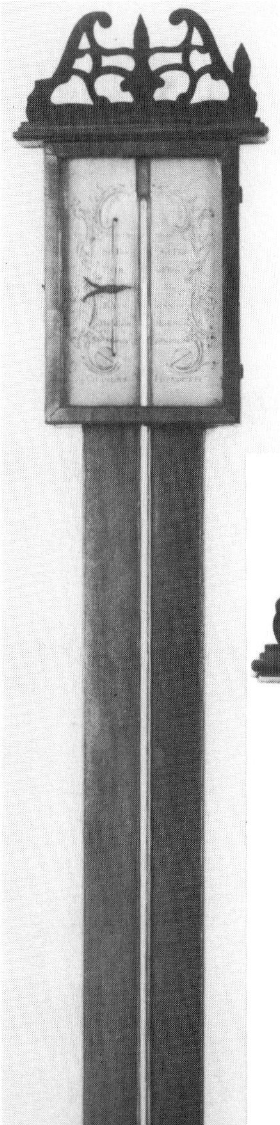

54

Set Fair
Set Frost
FAIR
RAIN
SNOW
Much Rain
Much Snow
Stormy
CHARLES
HOWORTH

55

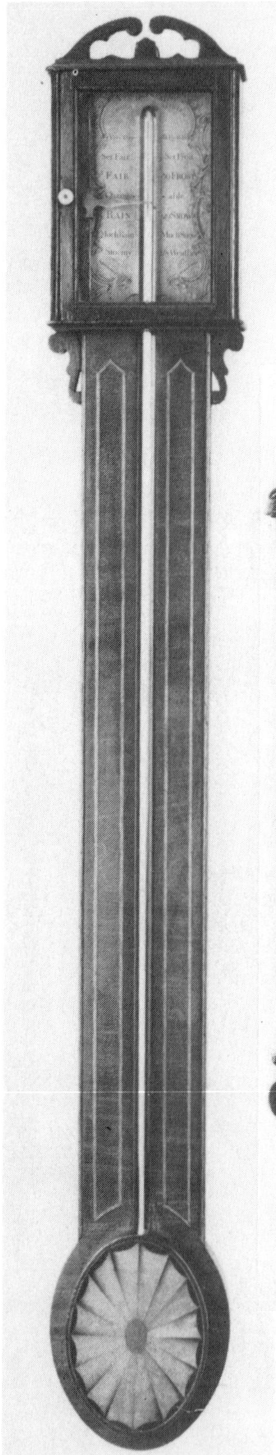

56

SUMMER
WINTER
Very Dry
31 Hard Frost
Set Fair
Set Frost
FAIR
30 FROST
Change
able
RAIN
29 SNOW
Much Rain
Much Snow
Stormy
28 Weather

57

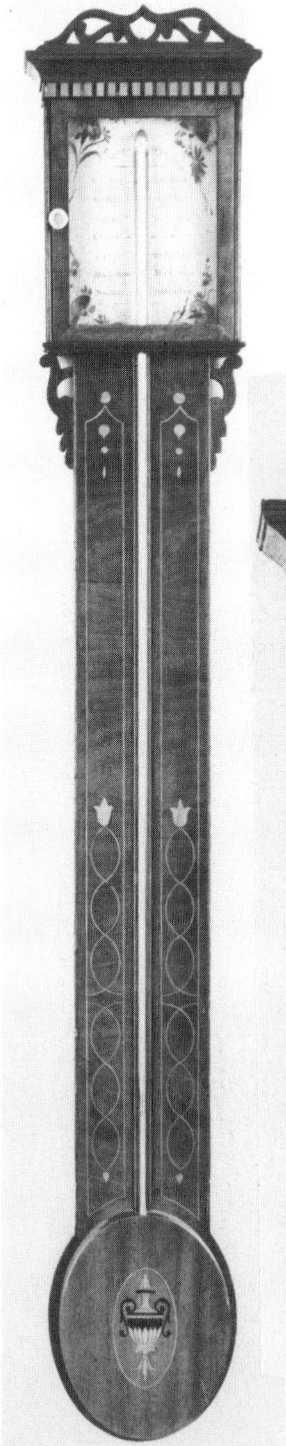

58

WINTER
Set Fair
Set Frost
Change
Much Rain
Much Snow
Stormy
28 Weather

59

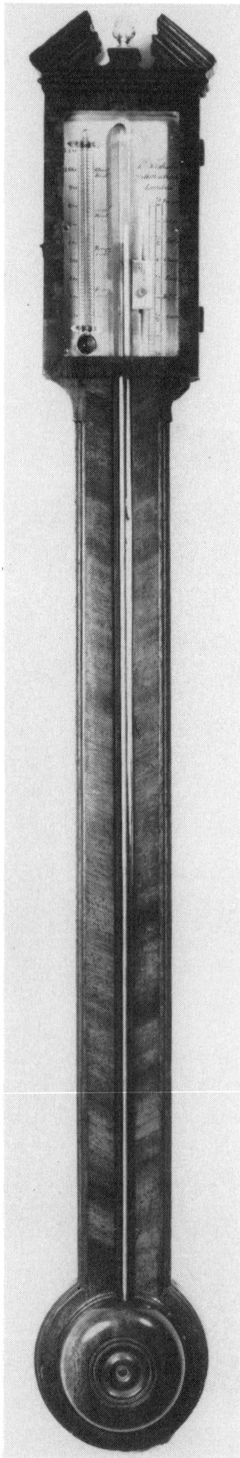

60

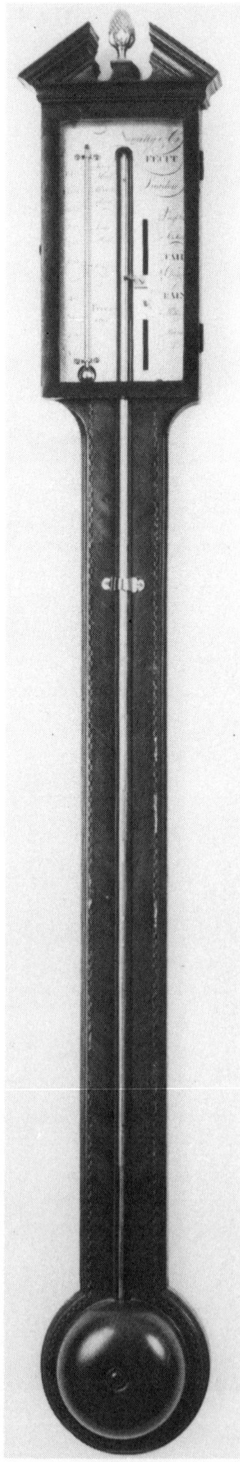

61

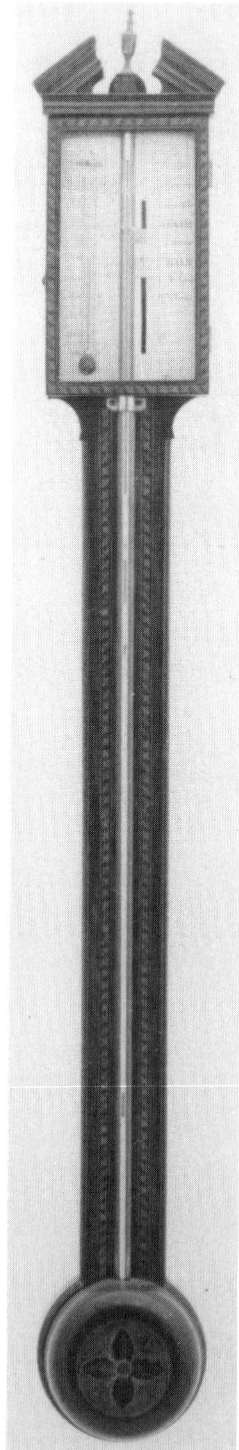

62

63

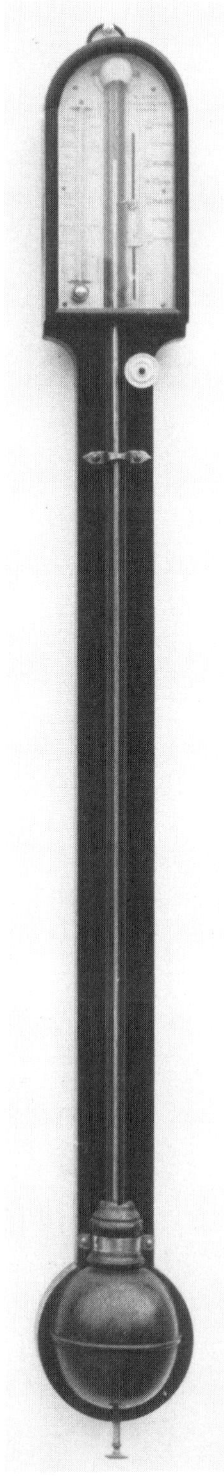
64

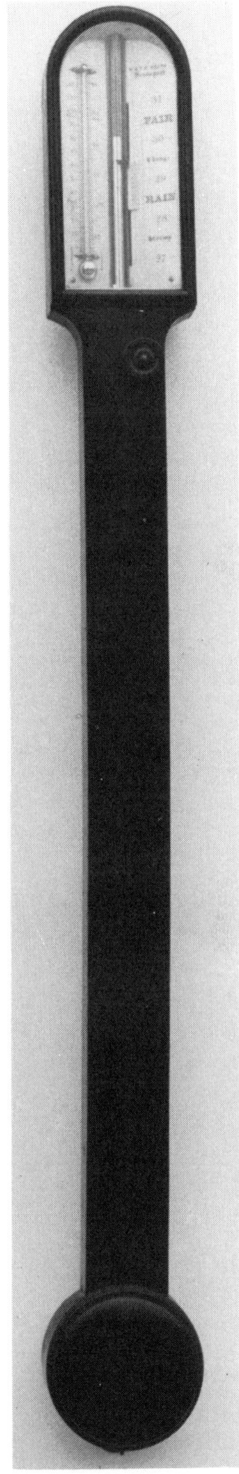

65

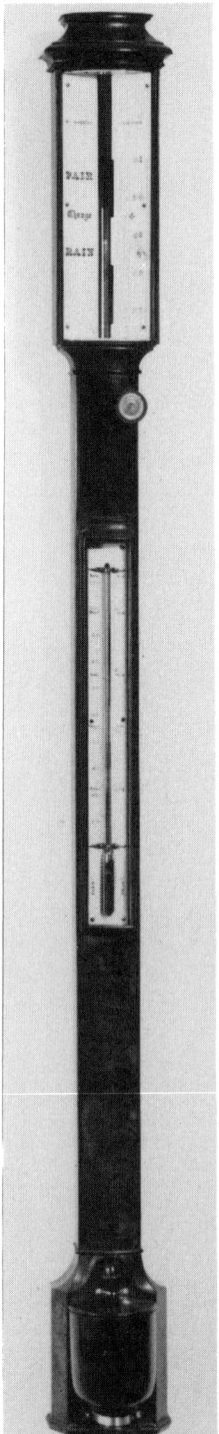
FAIR
Change
RAIN

66

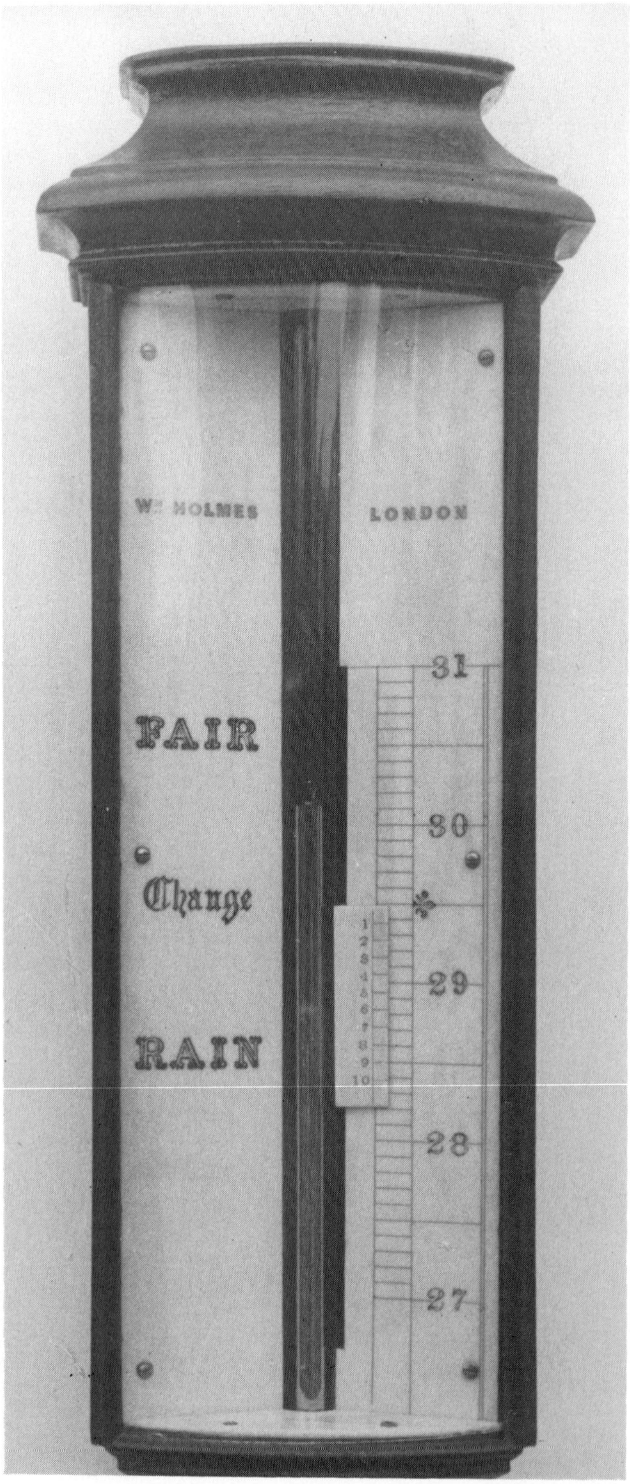
Wm HOLMES
LONDON
FAIR
Change
RAIN
31
30
29
28
27
1
2
3
4
5
6
7
8
9
10

67

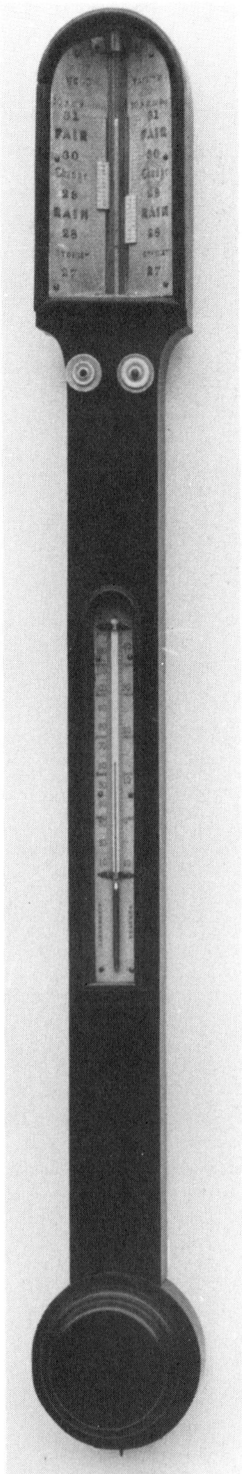

68

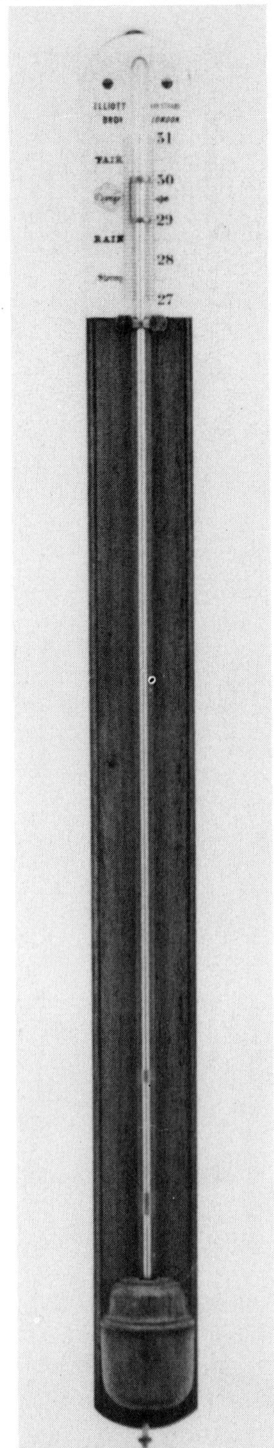

69

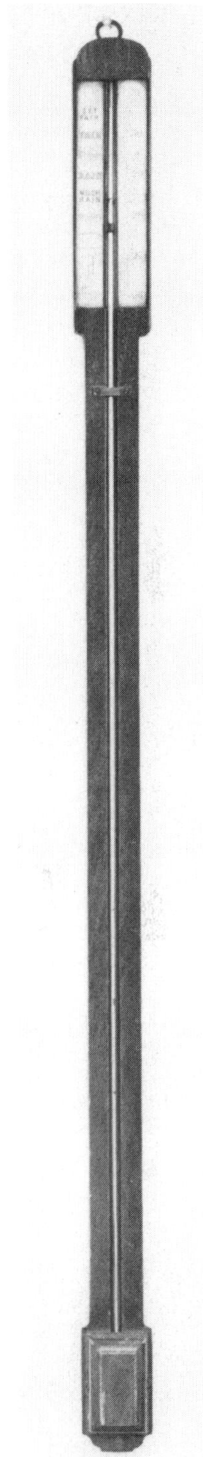

70

71

72

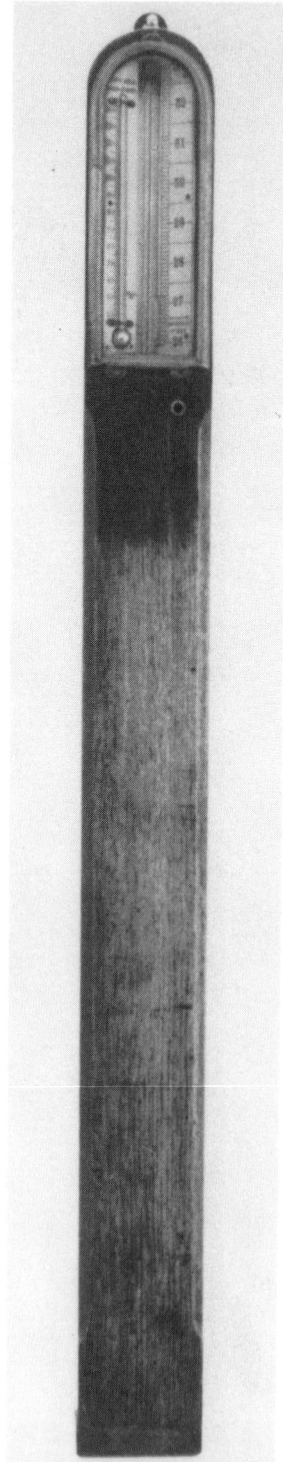

73

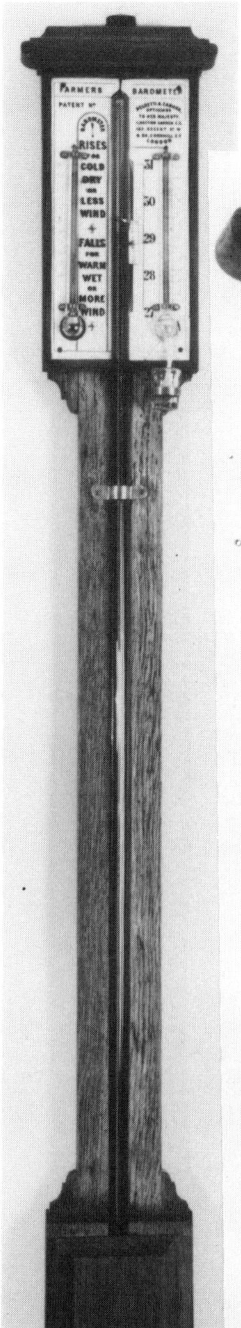

74

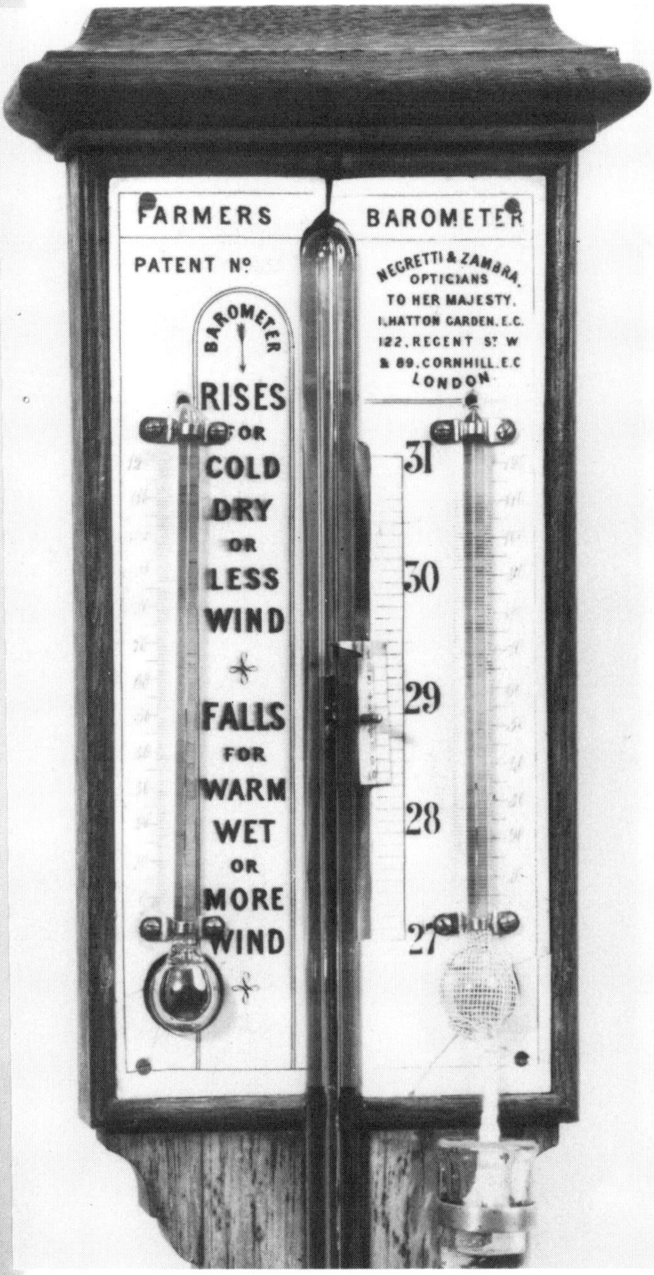
FARMERS
BAROMETER
PATENT No
NEGRETTI & ZAMBRA,
OPTICIANS
TO HER MAJESTY.
1, HATTON GARDEN, E.C.
122, REGENT St W
& 89, CORNHILL, E.C
LONDON
BAROMETER
RISES
FOR
COLD
DRY
OR
LESS
WIND
FALLS
FOR
WARM
WET
OR
MORE
WIND
31
30
29
28
27

75

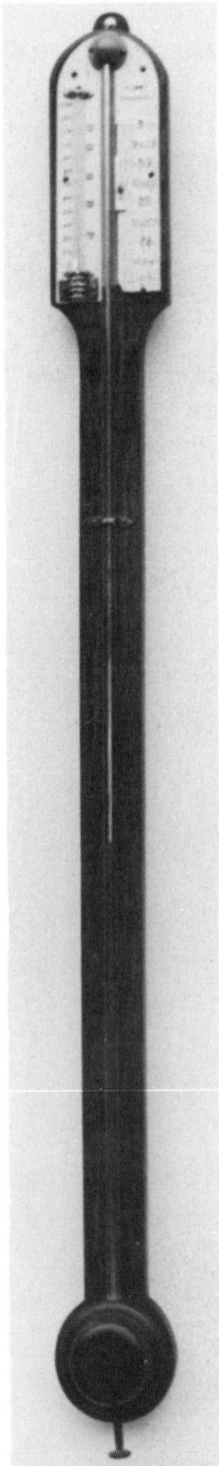

76

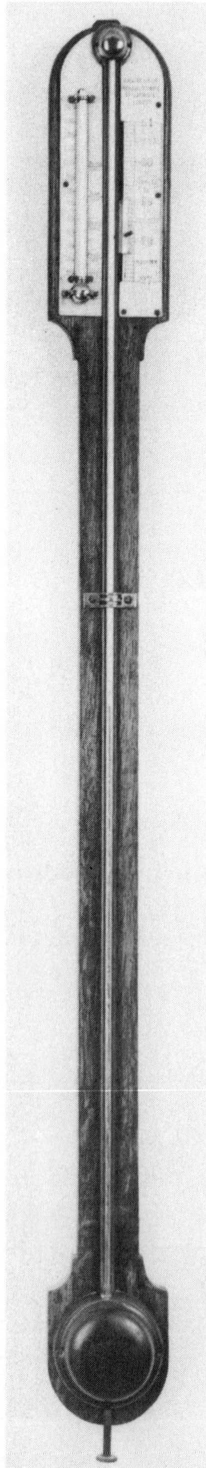

77

78

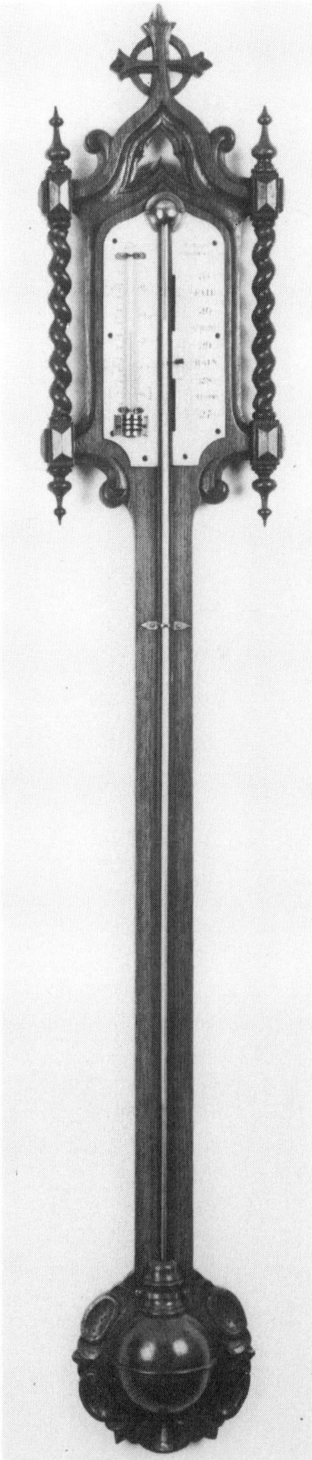

79

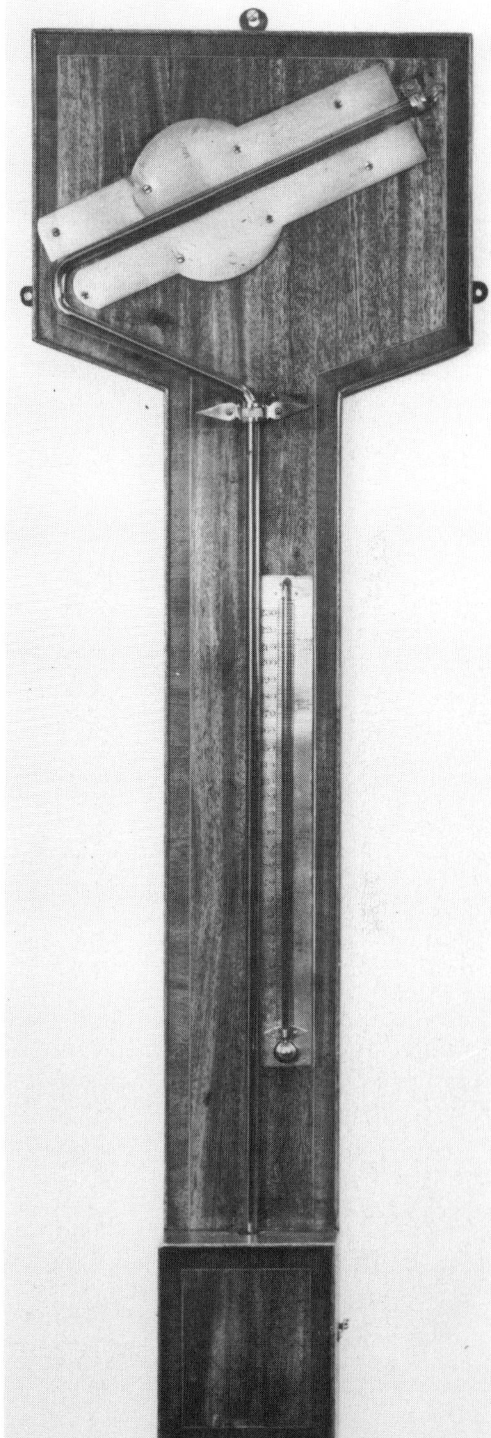

80

Stormy
Rain
Changeable
Fair
10
20
30
40
50

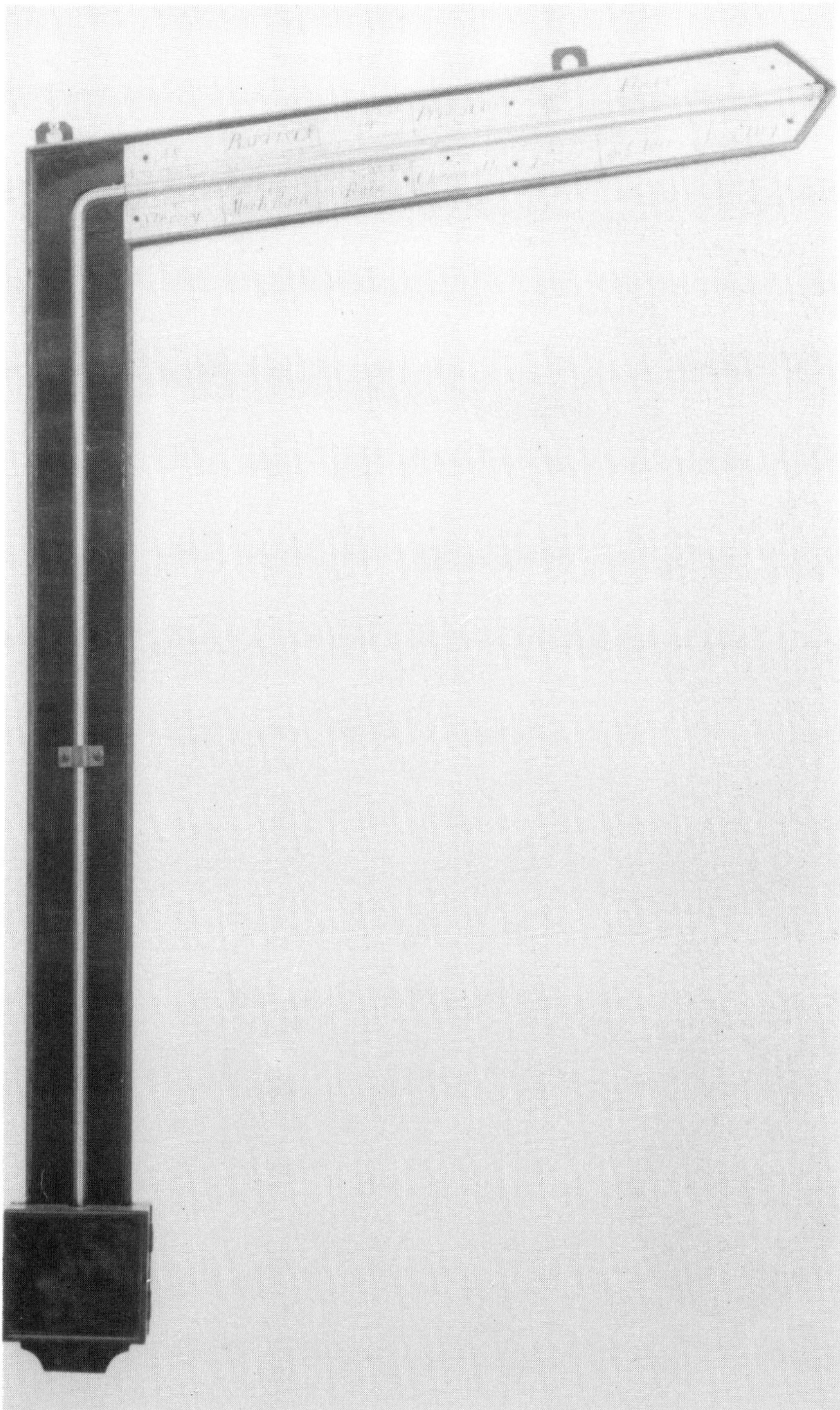

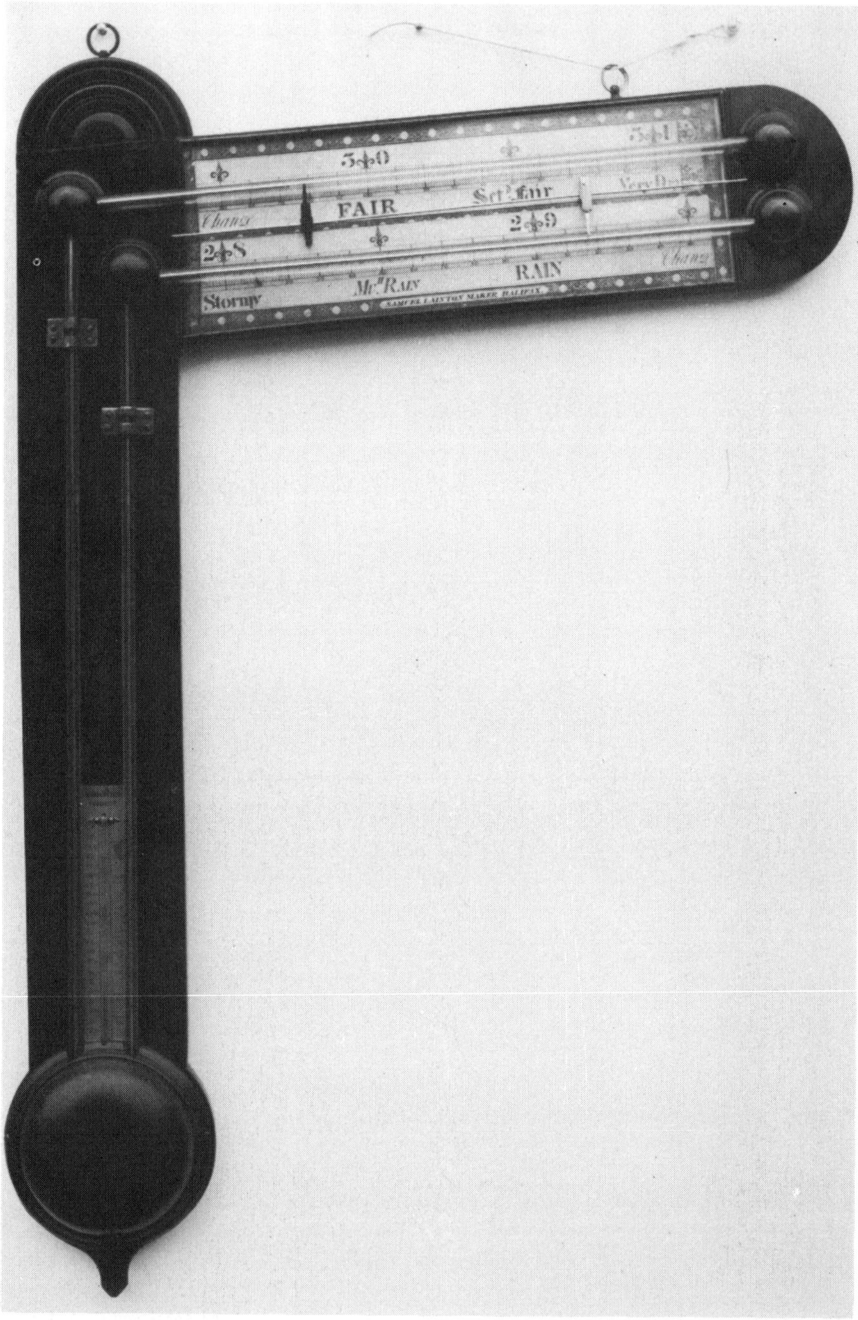
30
FAIR
29
28
RAIN
Stormy

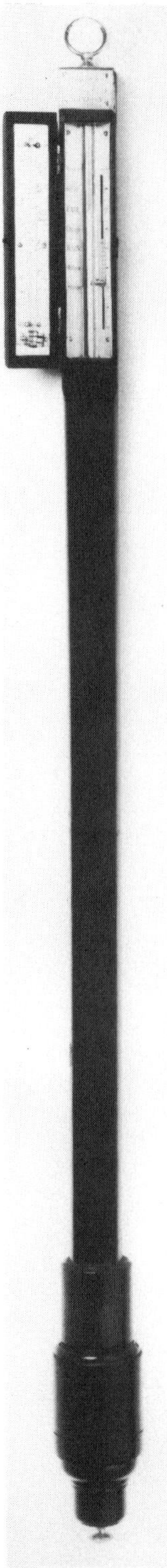

84

Thomas Jones
62
Charing Cross
LONDON
120
110
100
90
80
70
60
50
40
30
20
10
0
Fever Heat
Blood Heat
Sumr Heat
Temperate
Freezing
Very Dry
Set Fair
FAIR
Change
RAIN
Mh Rain
Stormy

85

86

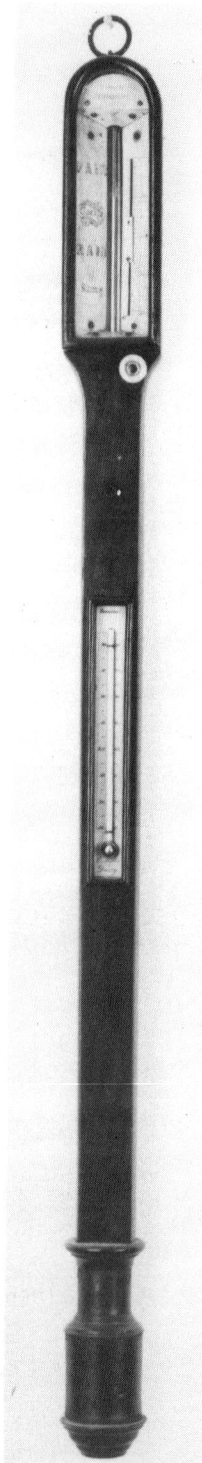

87

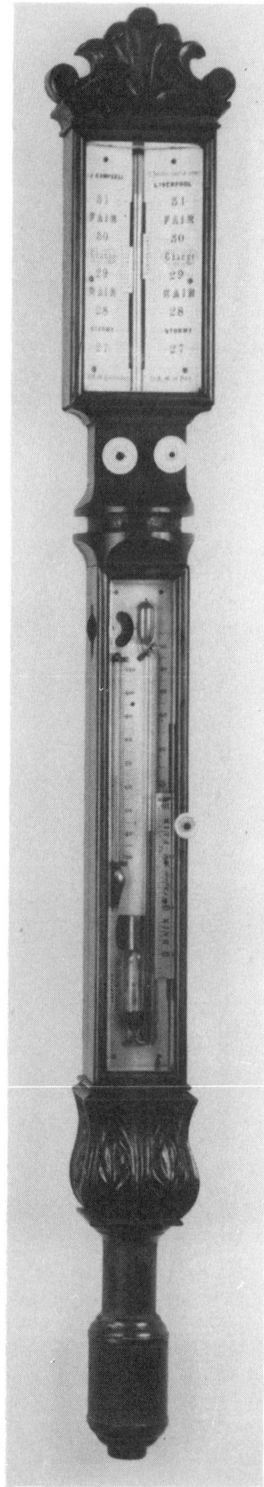

88

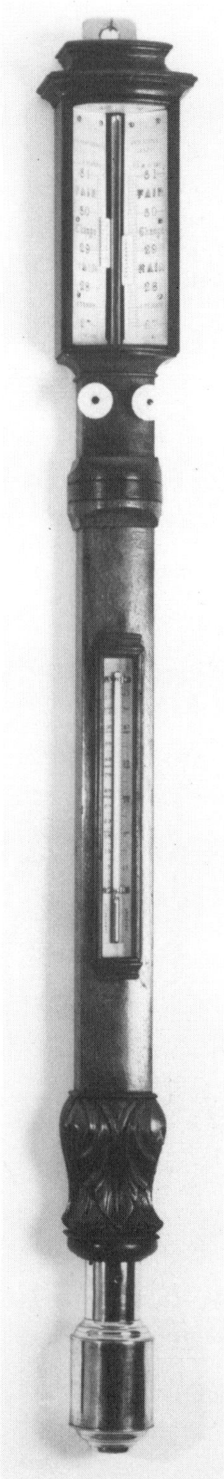

89

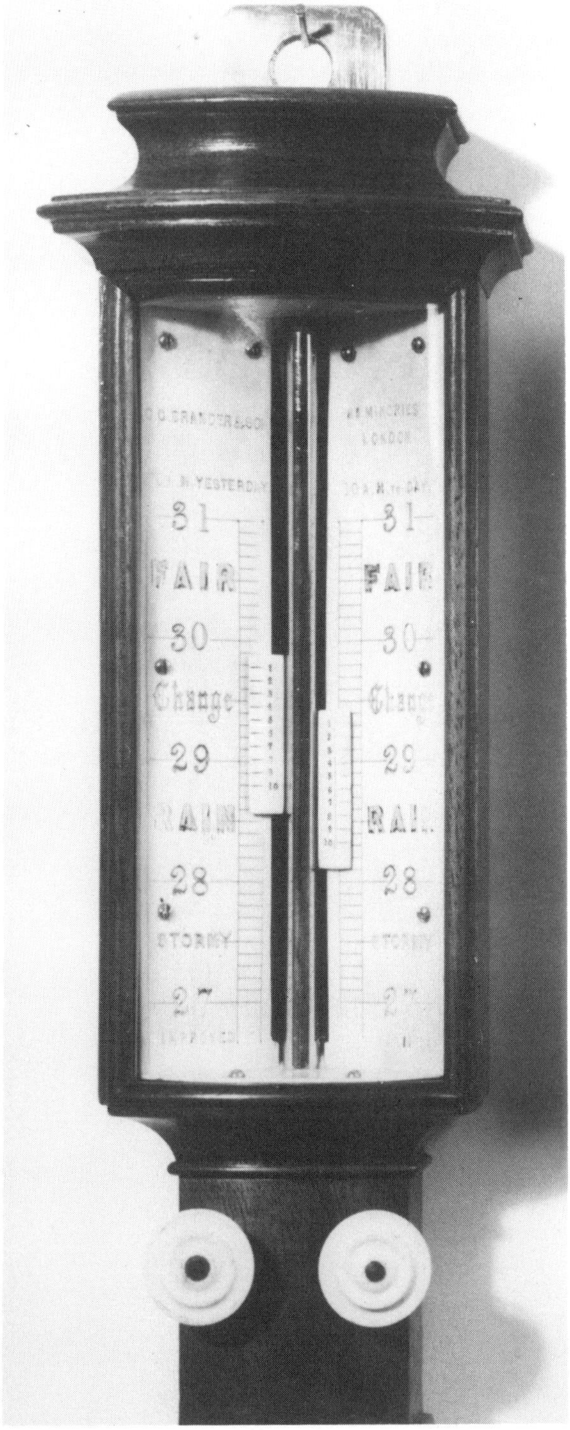

90

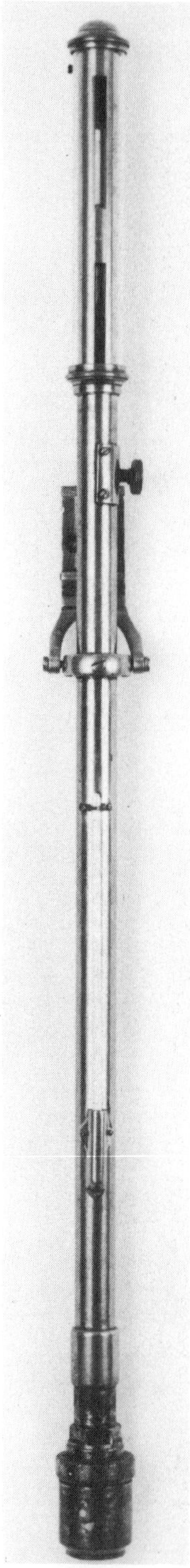

91

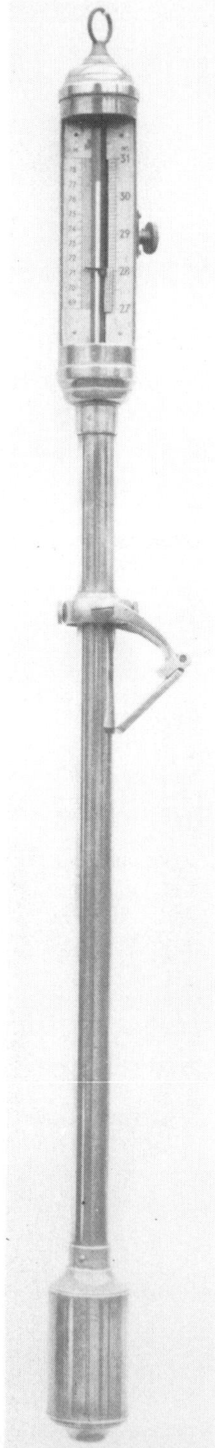

92

93

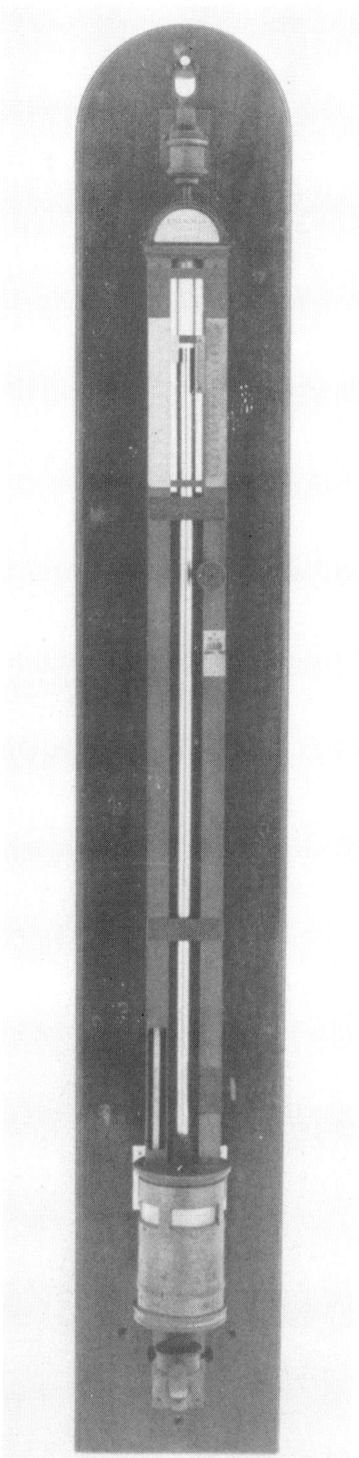

94

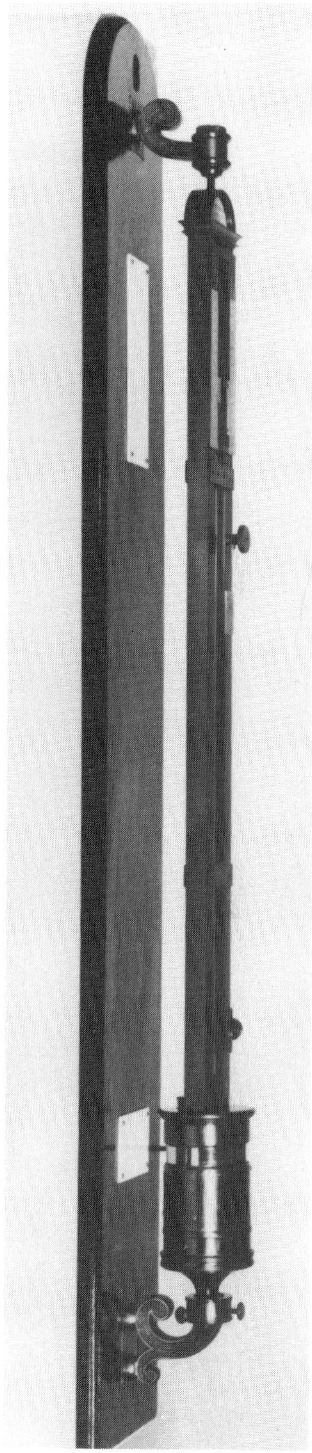

95

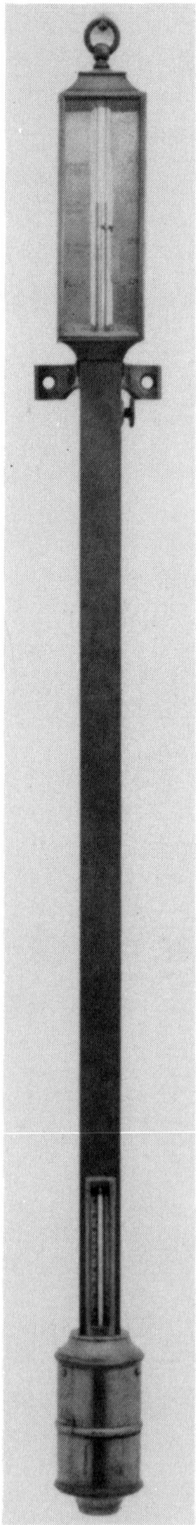

96

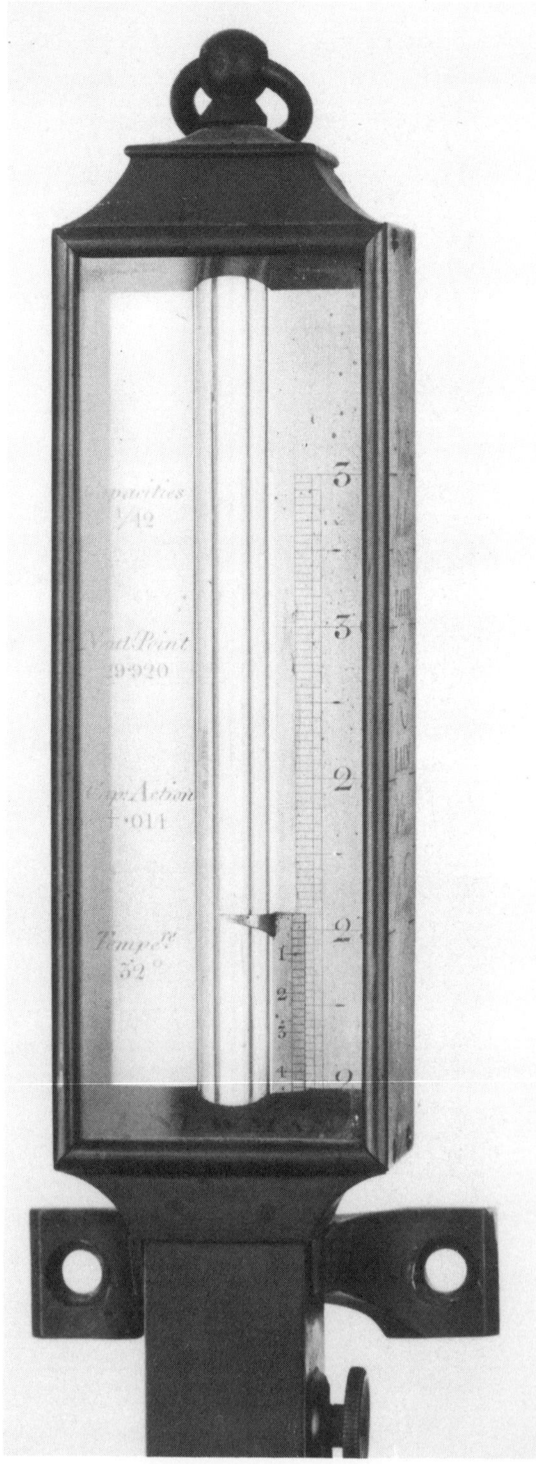

97

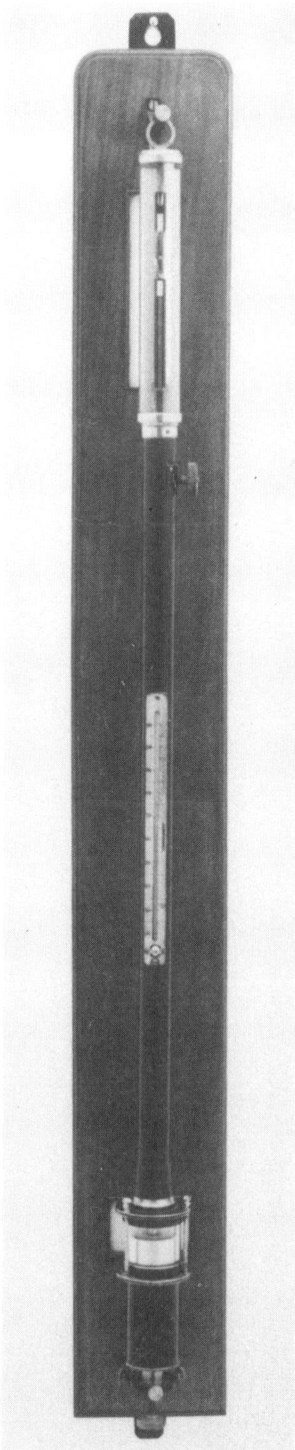
98

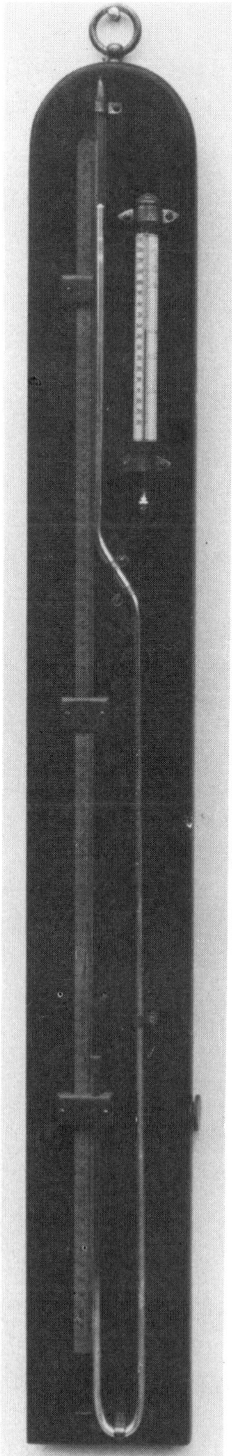
99

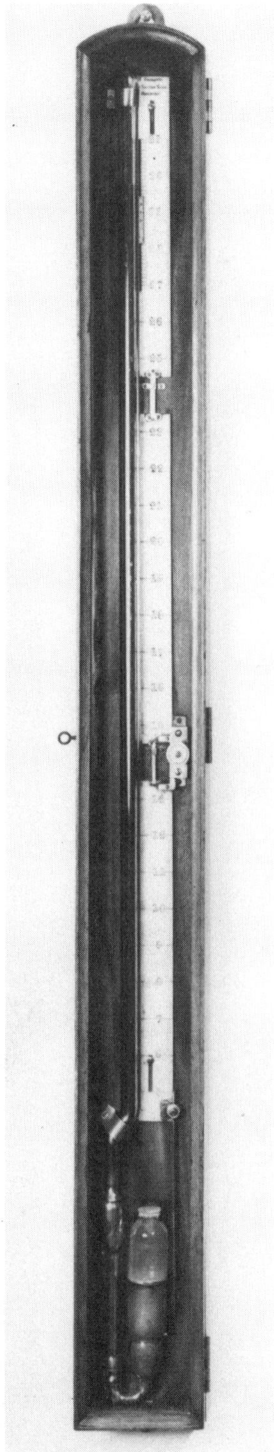
100

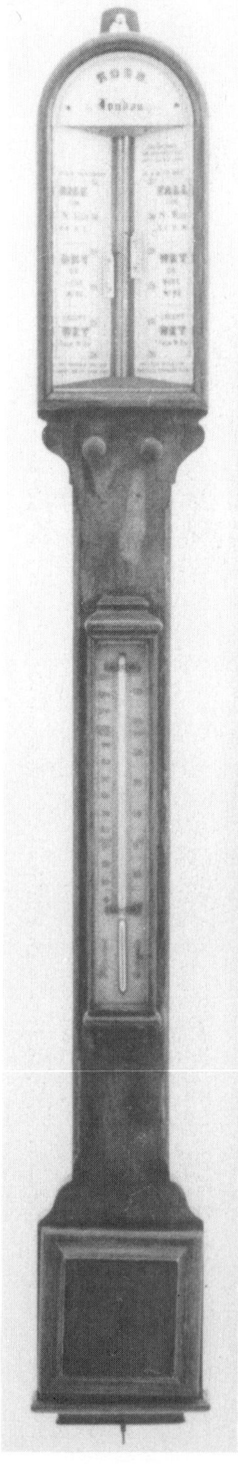

101

ROSS.
London
ADD ONE TENTH
FOR EACH NINETY FEET
ABOVE THE SEA LEVEL
10 A.M. YESTERDAY
10 A.M. TO DAY
31
30
29
28
27
26
RISE
FOR
N. ELY.
NW. N. E
DRY
OR
LESS
WIND
EXCEPT
WET
FROM N. Ed
FALL
FOR
S. WLY.
SE. S. W.
WET
OR
MORE
WIND
EXCEPT
WET
FROM N. Ed
LONG FORETOLD LONG LAST
SHORT NOTICE SOON PAST
FAST RISE AFTER LOW
FORETELLS STRONGER BLOW

102

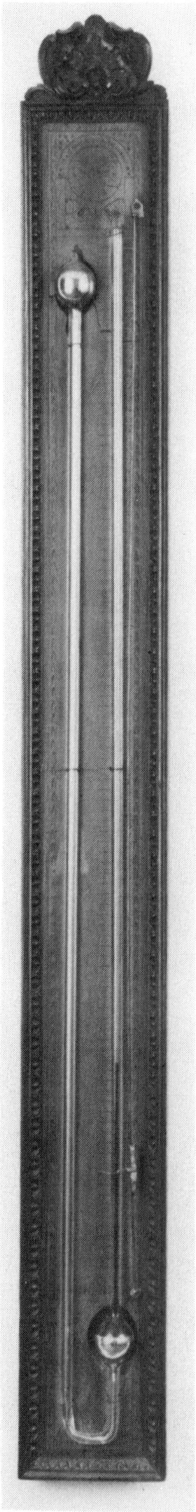

103

104

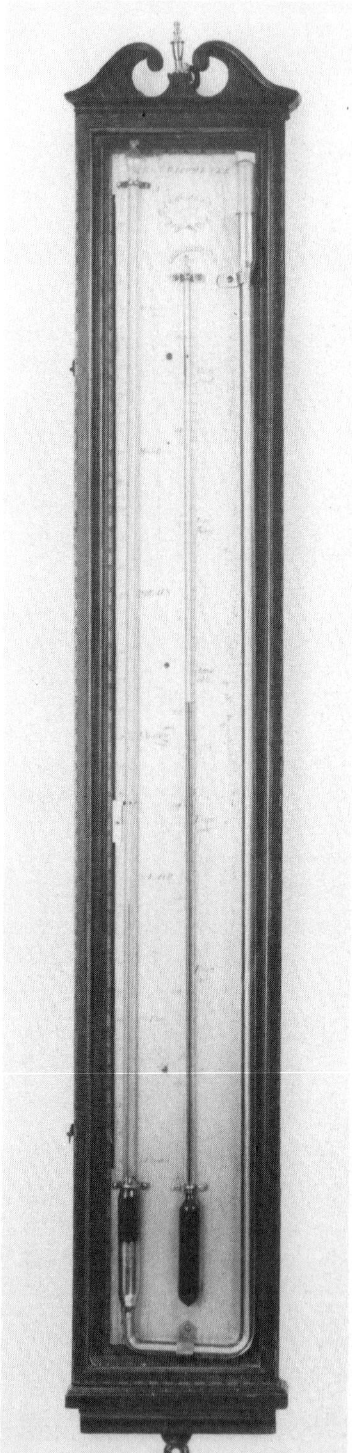

105

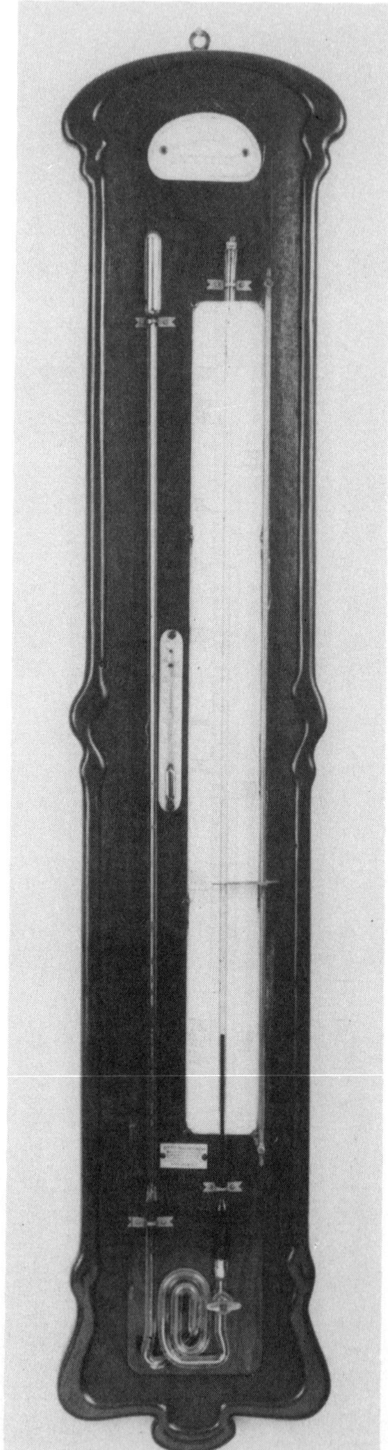

106

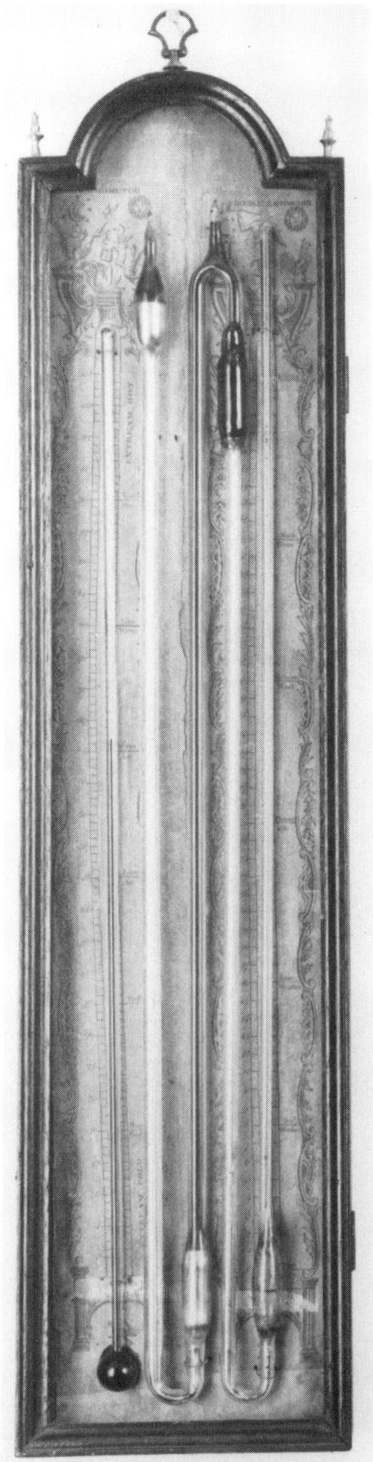

107

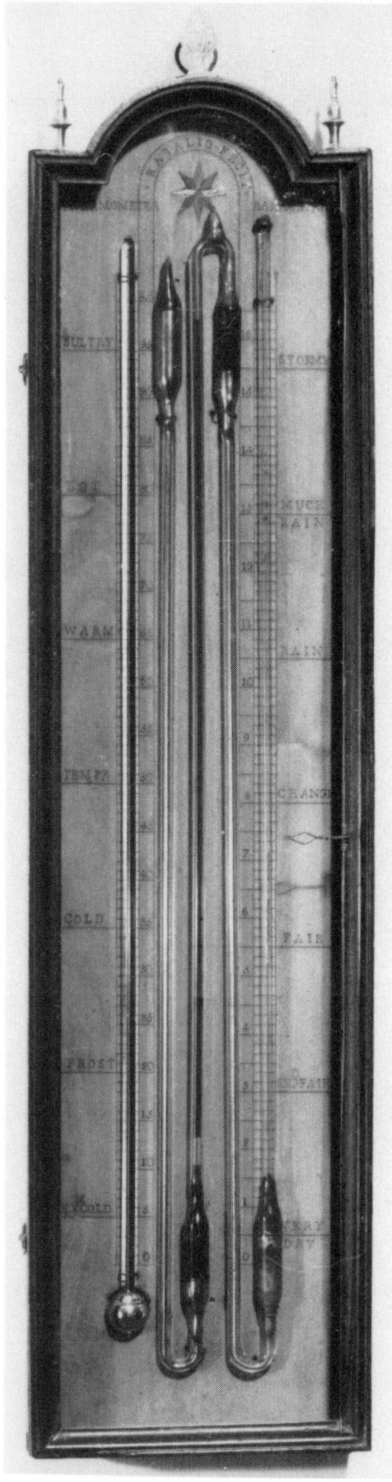

108

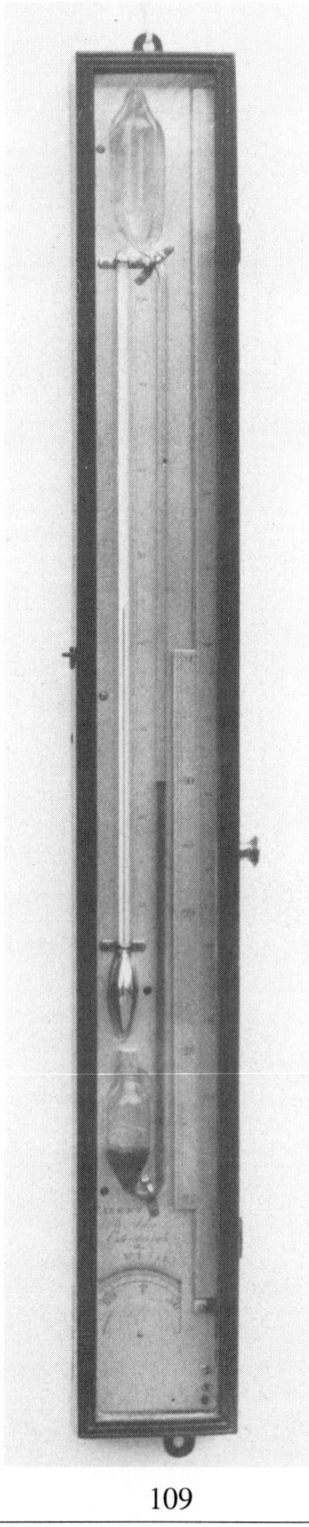

109

110

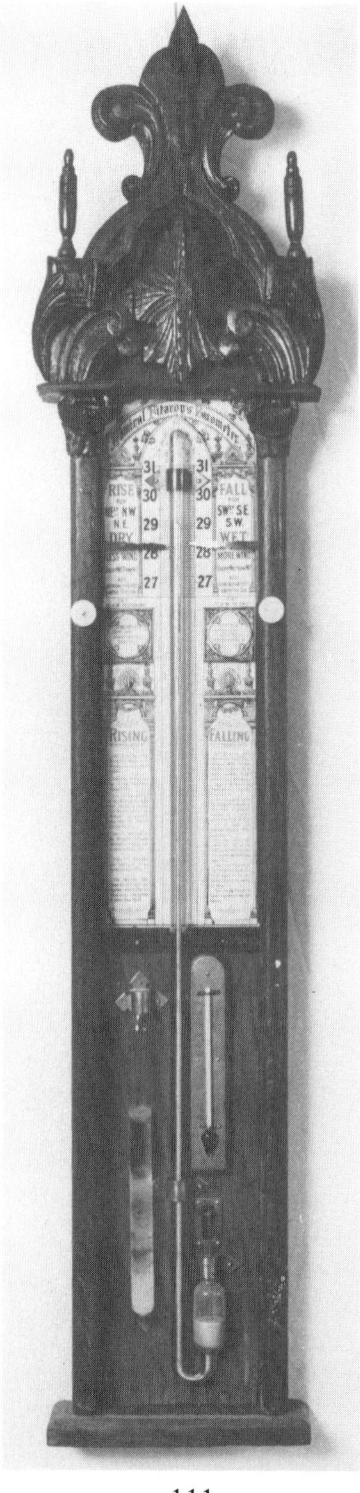
RISE
FALL
31
30
29
28
27
DRY
WET
RISING
FALLING

111

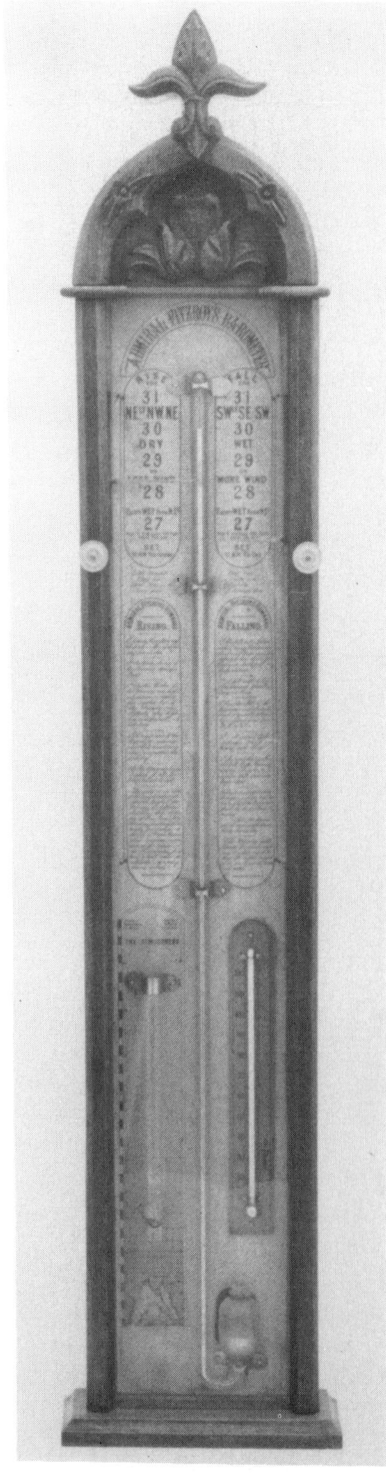
31
30
DRY
29
28
27
31
30
WET
29
MORE WIND
28
27
RISING
FALLING

112

BAROMETER RISES
BAROMETER FALLS
FOR N.E.LY
N.W. N.E.
DRY
OR
LESS WIND
EXCEPT WET
FOR S.WLY
S.E. S.W
WET
OR
MORE WIND
31
30
29
28
27
ADMIRAL FITZROY'S REMARKS

113

BAROMETER RISES
BAROMETER FALLS
FOR N.E.LY
N.W. N.E.
DRY
OR
LESS WIND
EXCEPT WET
FOR S.WLY
S.E. S.W
WET
OR
MORE WIND
31
30
29
28
27
ADMIRAL FITZROY'S REMARKS
MOUNTAIN SCALE
THE BAROMETER

114

ADMIRAL
FITZROYS BAROMETER
RISE
FALL
RISING
FALLING

115

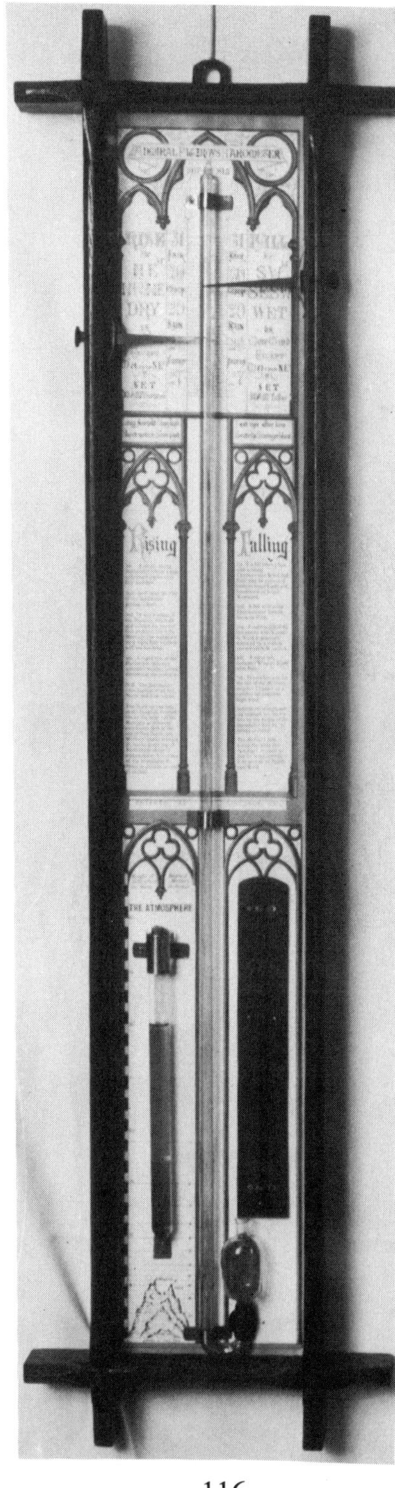
RISING
FALLING
THE ATMOSPHERE

116

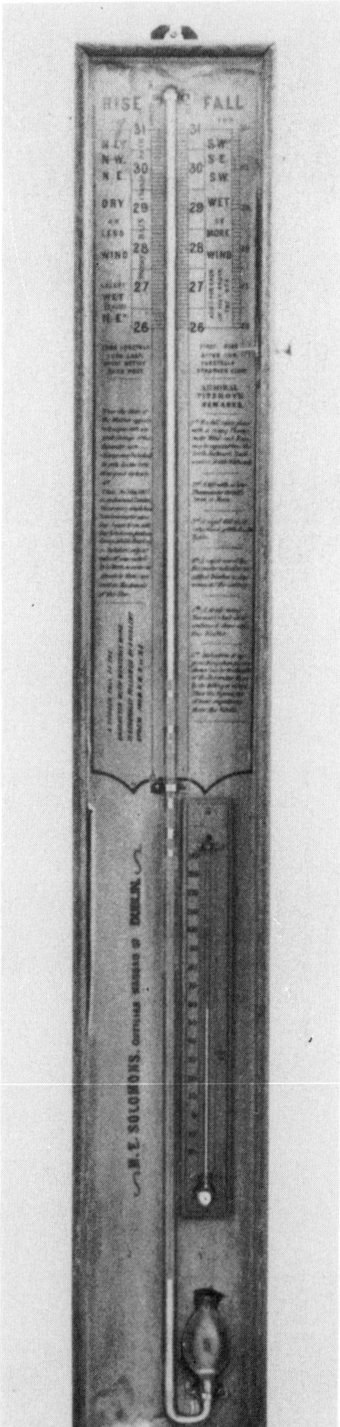

117

118

119

120

Changeable
RAIN
FAIR
29
30
Much Rain
Set Fair
28
31
Stormy
Very Dry

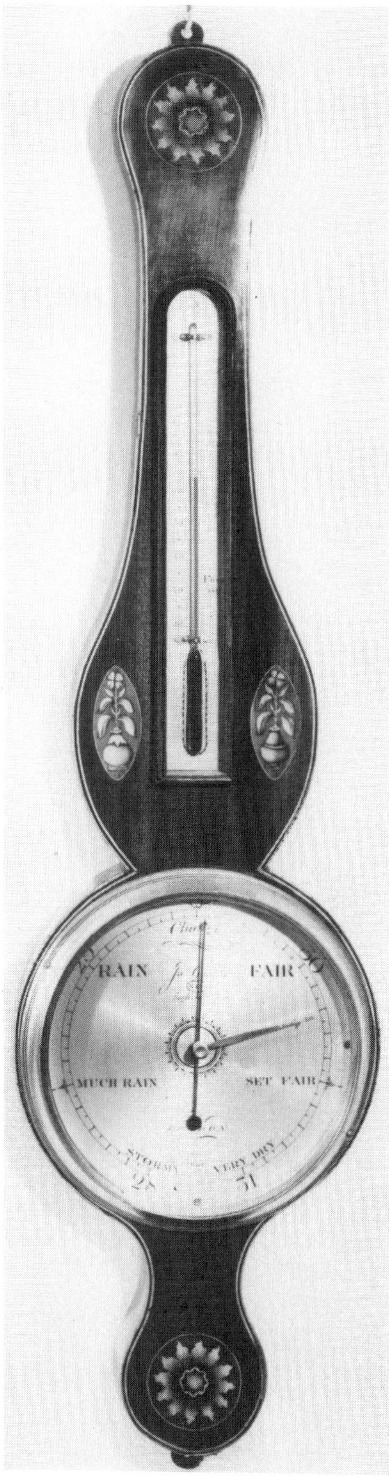
RAIN
FAIR
MUCH RAIN
SET FAIR
STORMY
VERY DRY
31
122

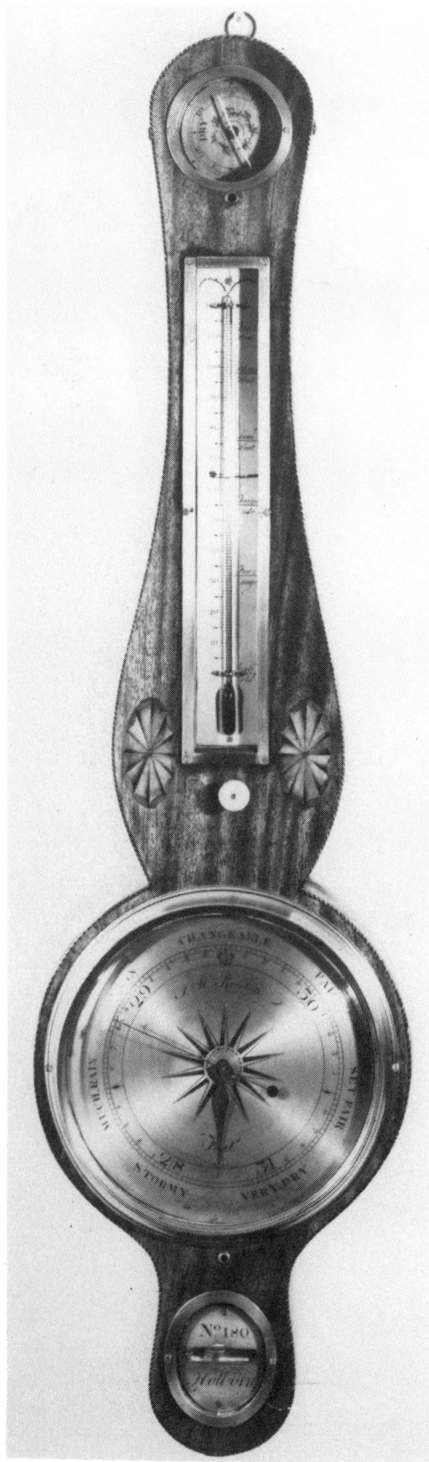
CHANGEABLE
29
30
MUCH RAIN
SET FAIR
28
31
STORMY
VERY DRY
No 180
123

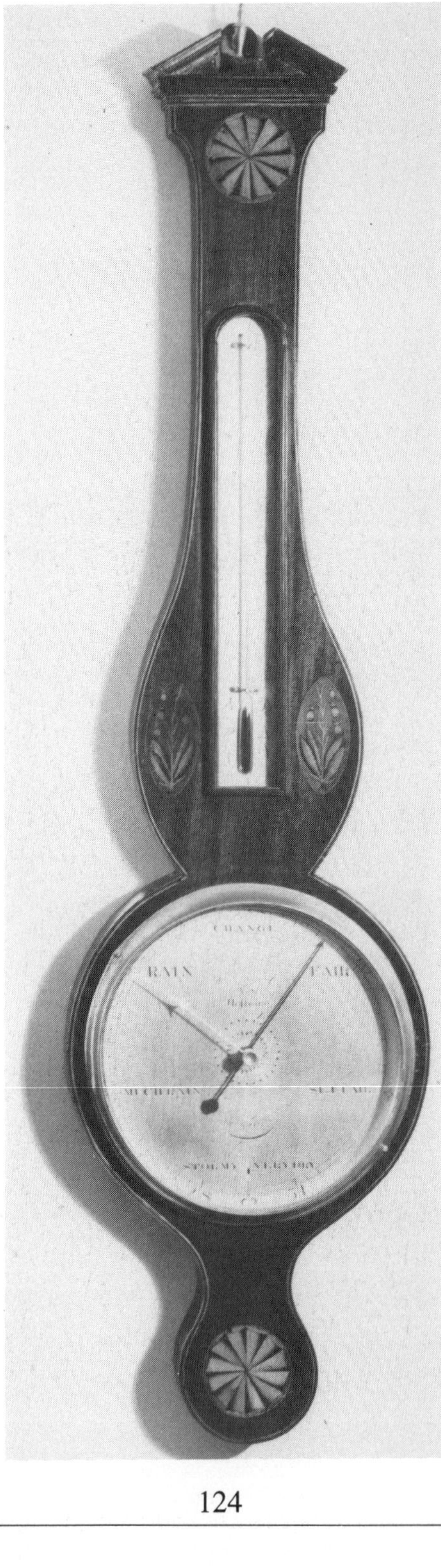

124

125

Changeable
RAIN
FAIR
29
30
P. Gally
MUCH RAIN
SET FAIR
London
28
31
STORMY
VERY DRY

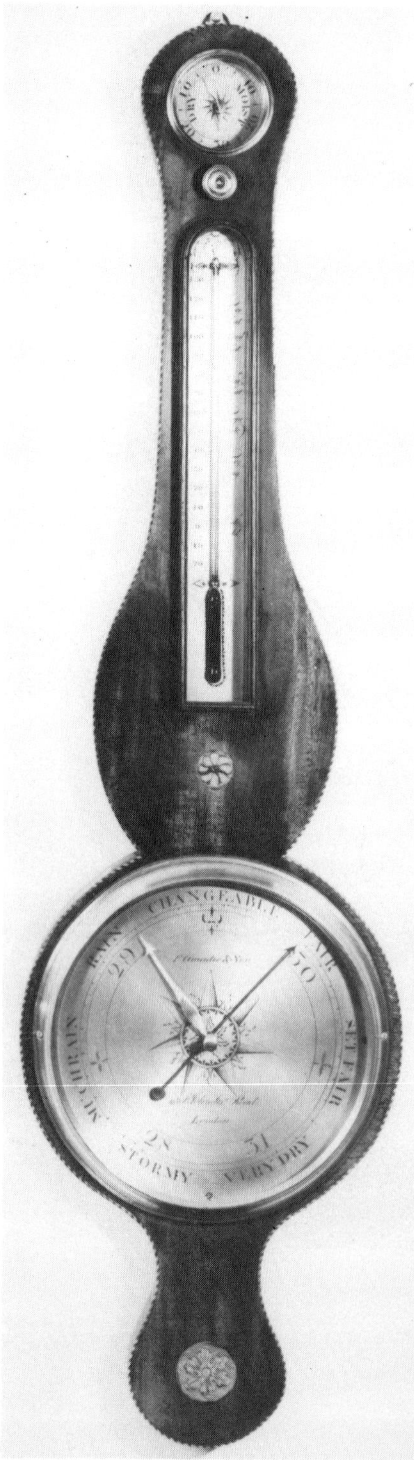
CHANGEABLE
RAIN
FAIR
29
30
MUCH RAIN
SET FAIR
28
31
STORMY
VERY DRY

127

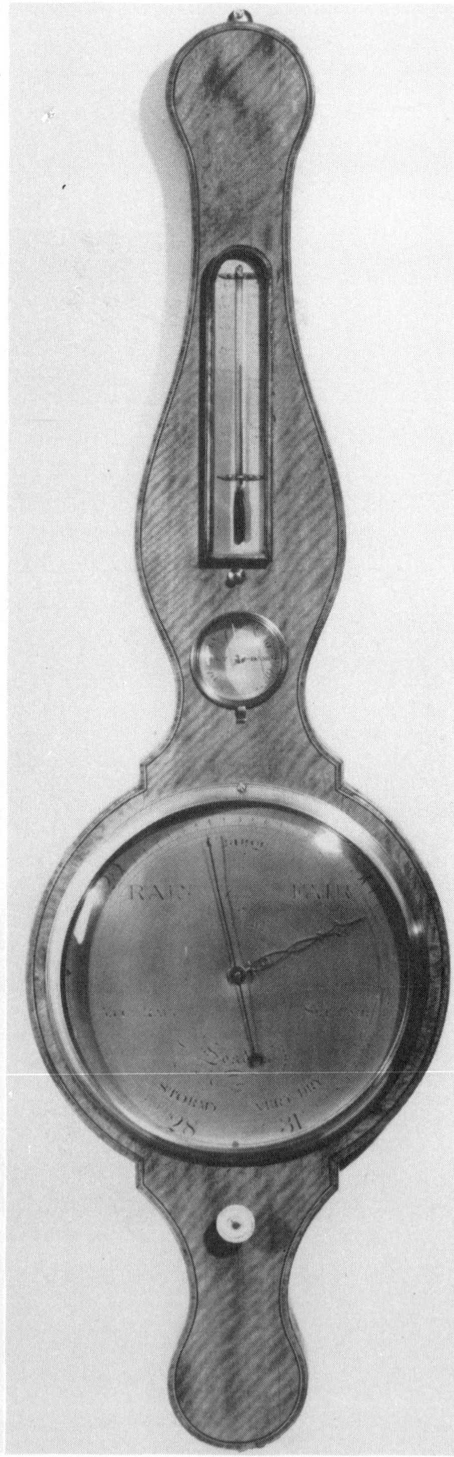
RAIN
FAIR
STORMY
VERY DRY
28
31

128

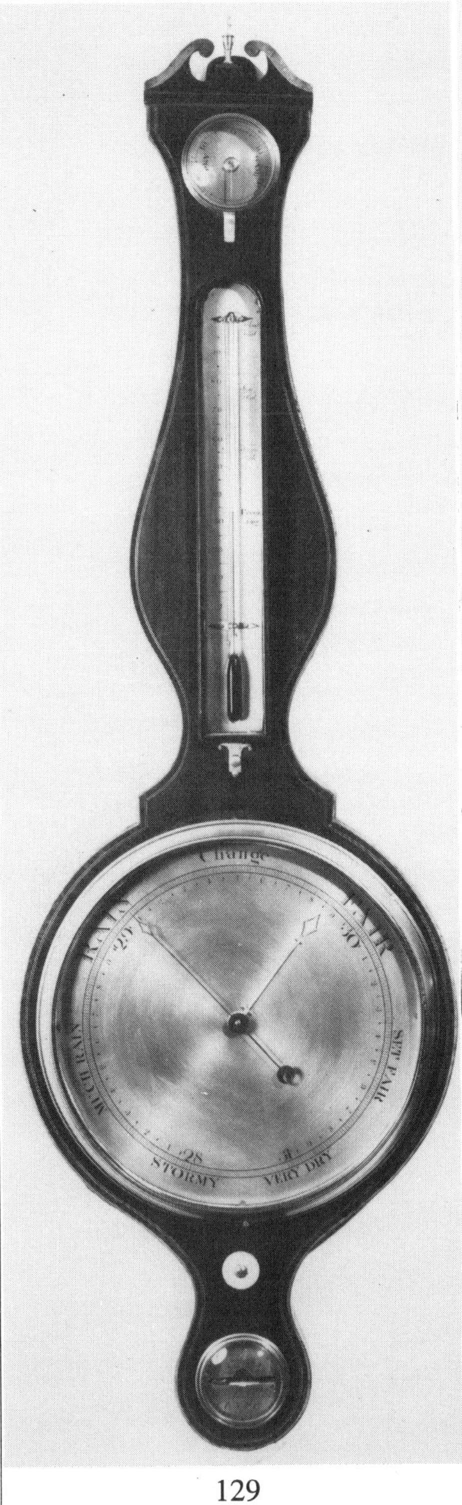
Change
RAIN
FAIR
29
30
MUCH RAIN
SET FAIR
28
STORMY
VERY DRY

129

Change
RAIN
FAIR
MUCH RAIN
SET FAIR
STORMY
VERY DRY
28
31

130

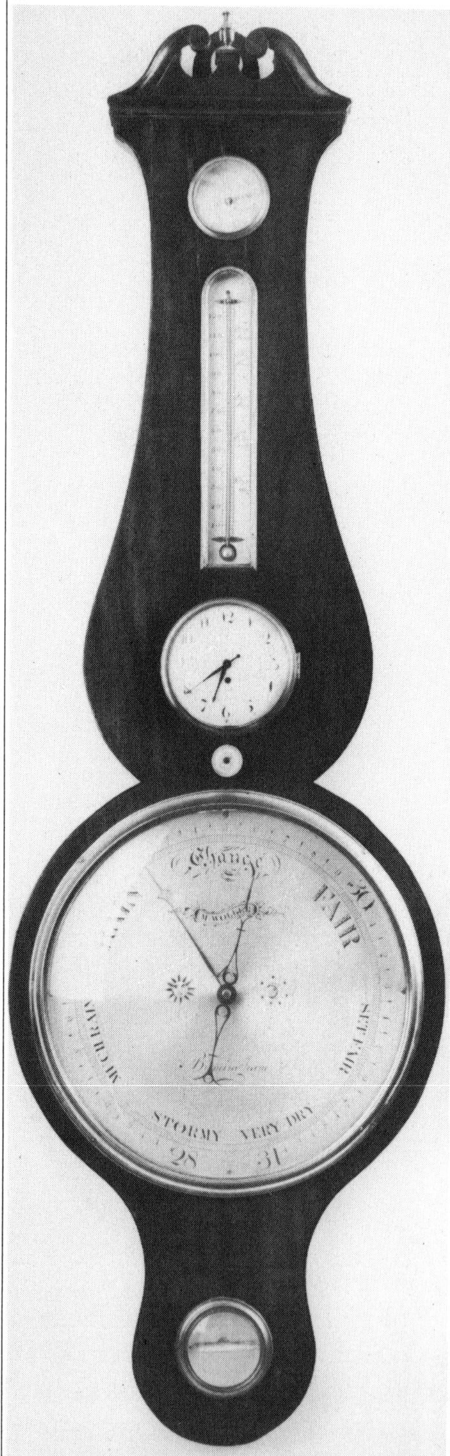
Change
FAIR
30
STORMY VERY DRY
28
31

131

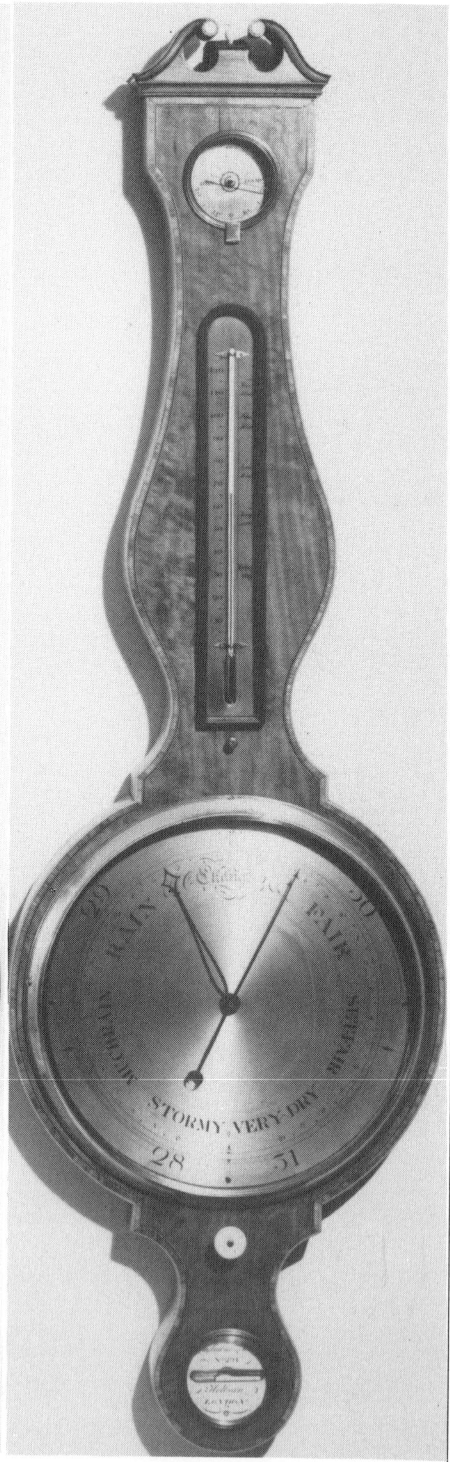
29
RAIN
Change
FAIR
30
STORMY, VERY DRY
28
31

132

RAIN
FAIR
30
SET FAIR
STORMY
VERY DRY
31
28

133

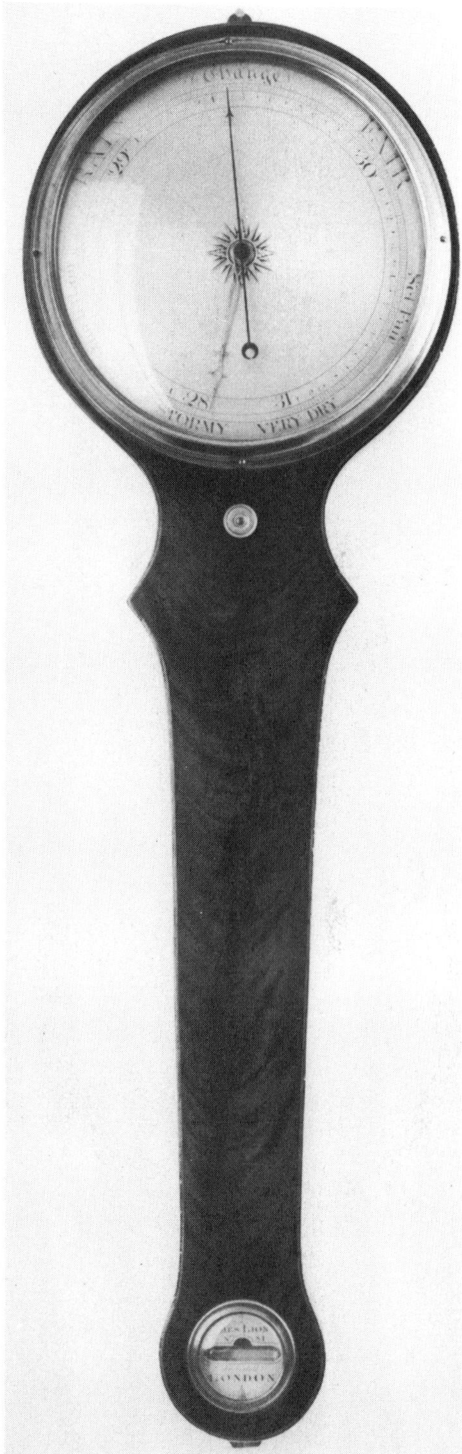
FAIR
30
31
VERY DRY
LONDON

134

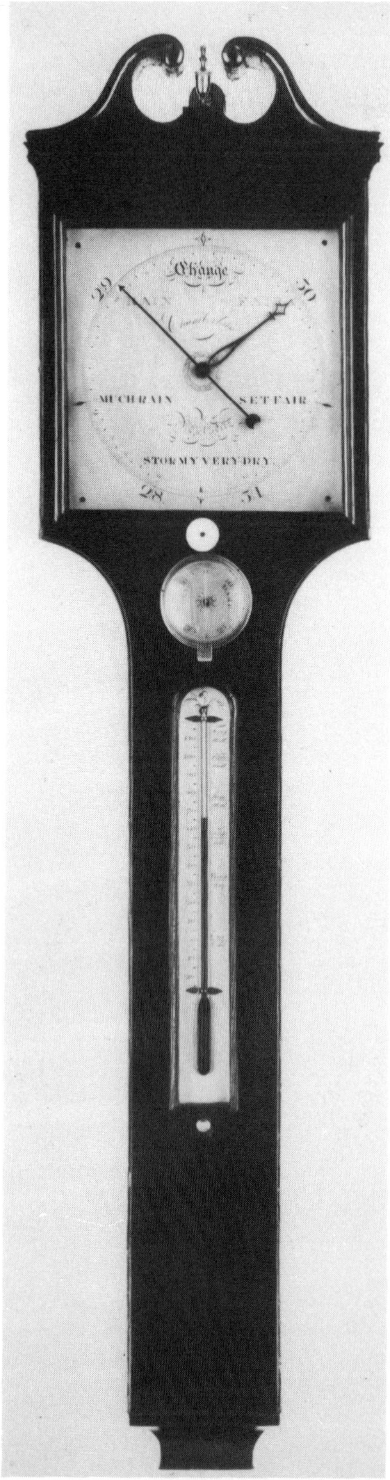
Change
29
30
MUCH RAIN
SET FAIR
STORMY VERY DRY
28
31
135

Change
RAIN
FAIR
STORMY
VERY DRY
136

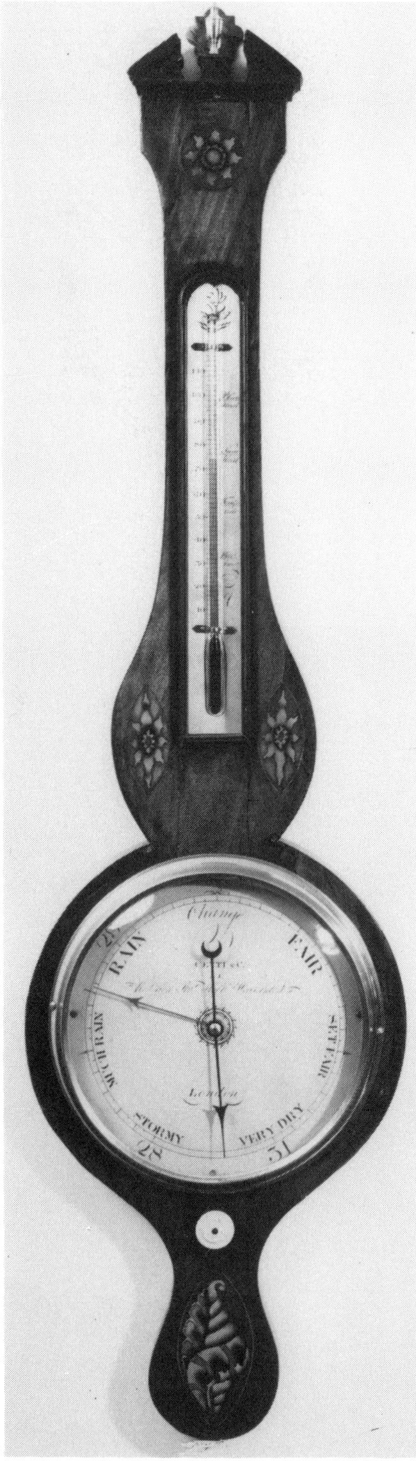
Change
RAIN
FAIR
MUCH RAIN
SET FAIR
London
STORMY
VERY DRY
28
31

137

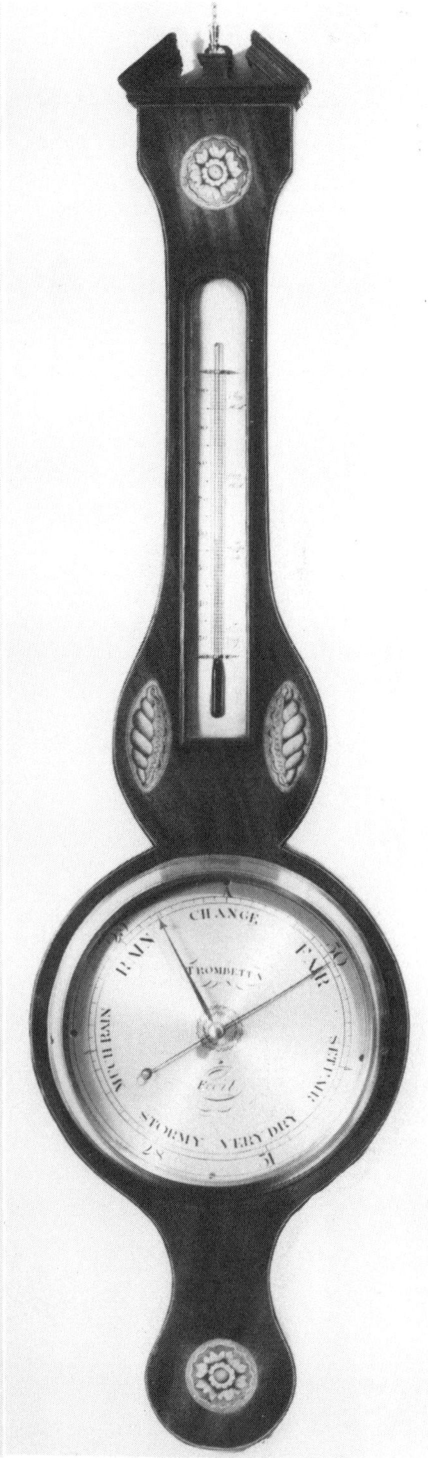
CHANGE
RAIN
FAIR
30
TROMBETTA
MUCH RAIN
SET FAIR
STORMY
VERY DRY
28
31

138

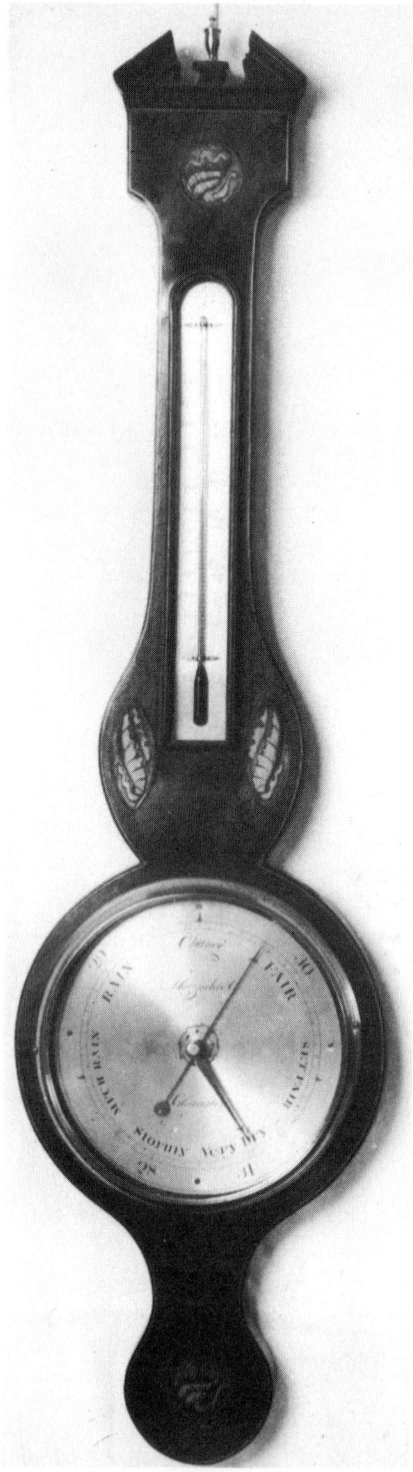
Change
29
Rain
Fair
30
Much Rain
Set Fair
Stormy
Very Dry
28
31

139

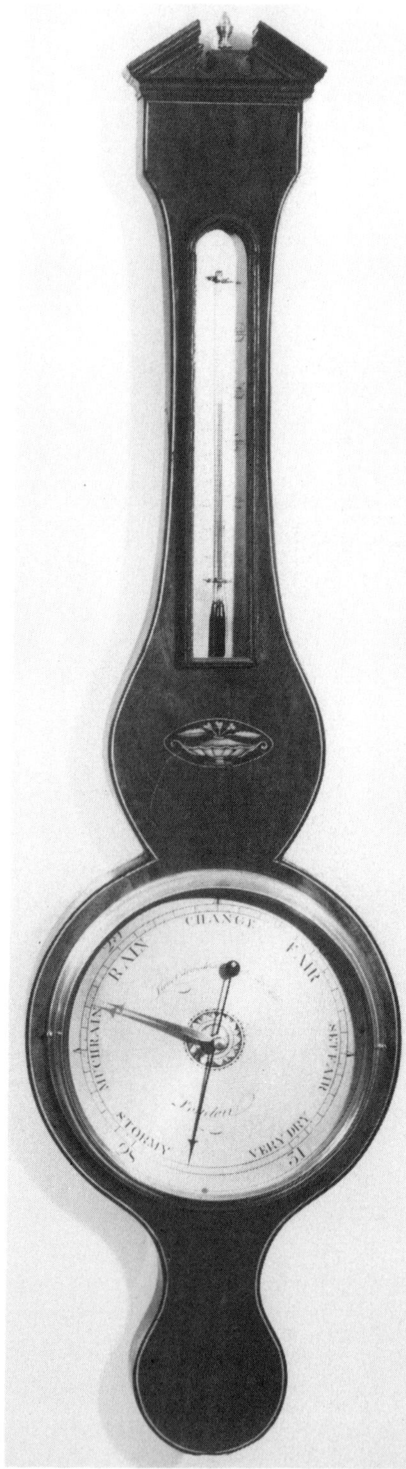
CHANGE
29
RAIN
FAIR
MUCH RAIN
SET FAIR
London
STORMY
VERY DRY
28
31

140

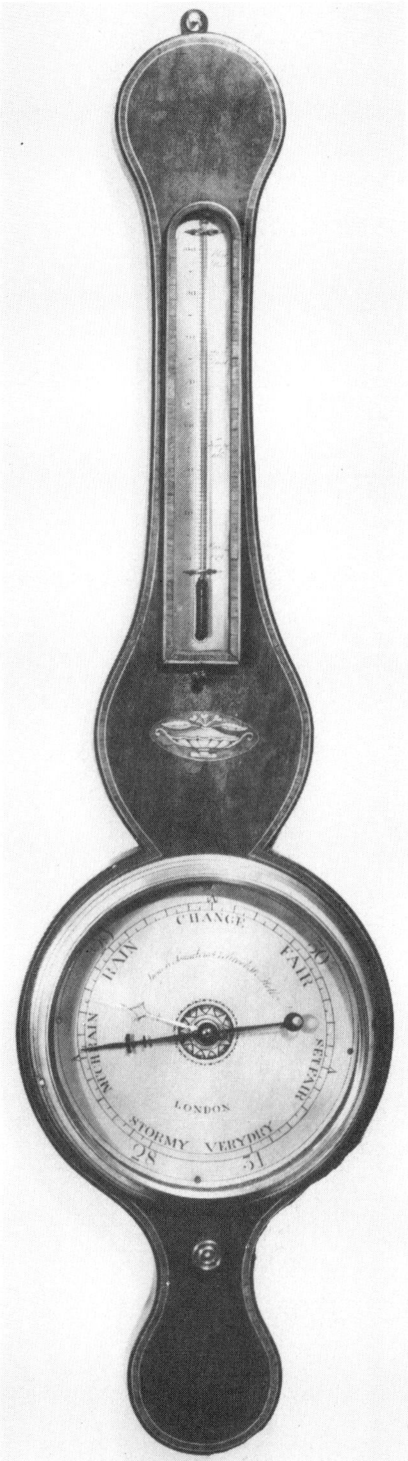
CHANGE
RAIN
FAIR
30
MUCH RAIN
SET FAIR
LONDON
STORMY
VERYDRY
28
31
141

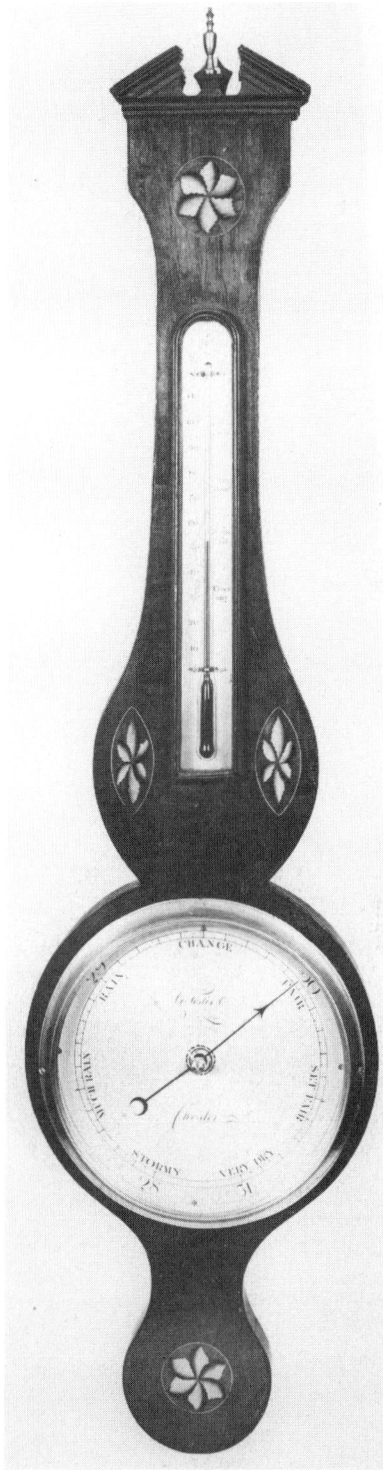
CHANGE
RAIN
FAIR
30
MUCH RAIN
SET FAIR
STORMY
VERY DRY
28
31
142

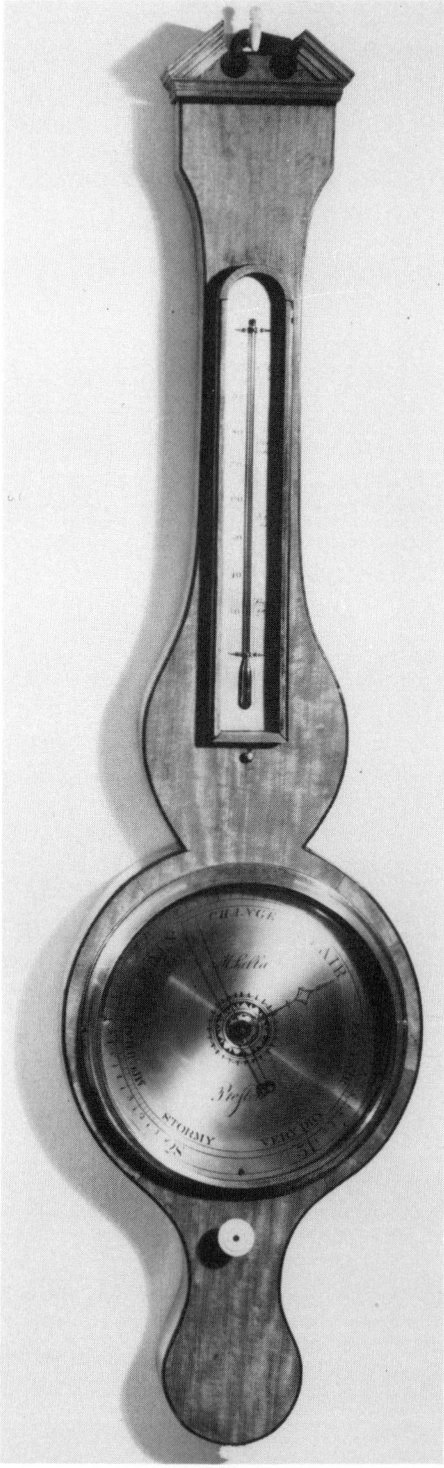
CHANGE
FAIR
STORMY
VERY DRY
143

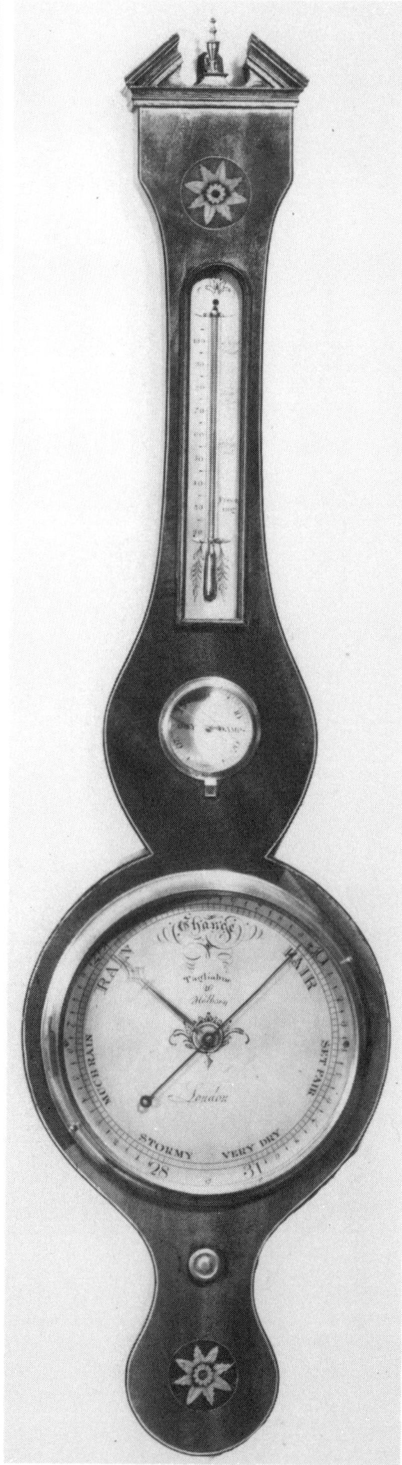
Change
RAIN
FAIR
MUCH RAIN
SET FAIR
London
STORMY
VERY DRY
144

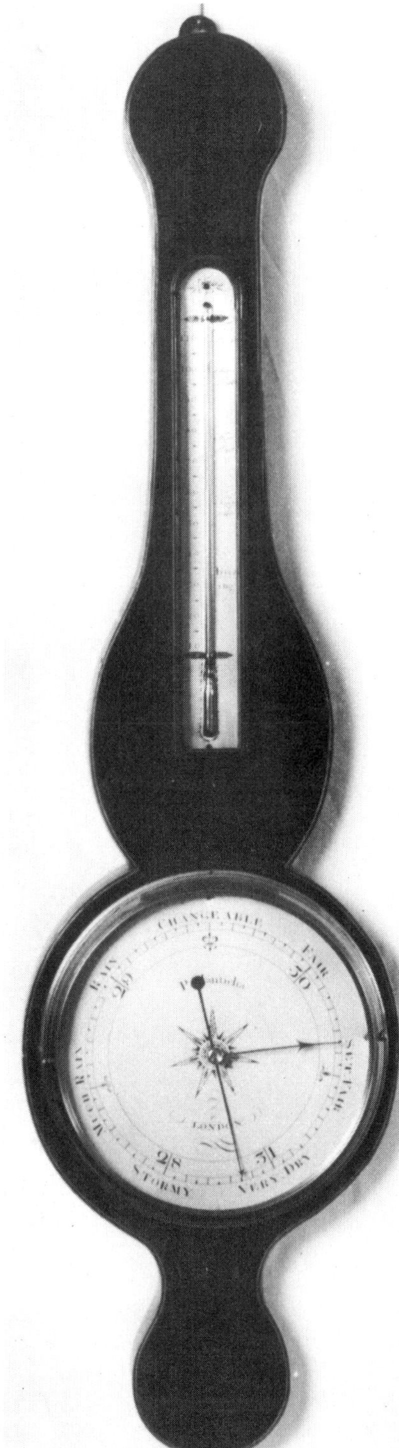
CHANGEABLE
RAIN
FAIR
29
30
MUCH RAIN
SET FAIR
LONDON
28
31
STORMY
VERY DRY

145

Change
RAIN
FAIR
29
30
MUCH RAIN
SET FAIR
STORMY
VERY DRY
28
31

146

Change
29
RAIN
30
FAIR
MUCH RAIN
SET FAIR
STORMY
VERY DRY
28
31

CHANGE
FAIR
RAIN
MUCH RAIN
SET FAIR
STORMY
VERY DRY
31
148

CHANGE
RAIN
FAIR
29
30
F Molton
Norwich
Mh RAIN
St FAIR
28
31
STORMY
VERY DRY
149

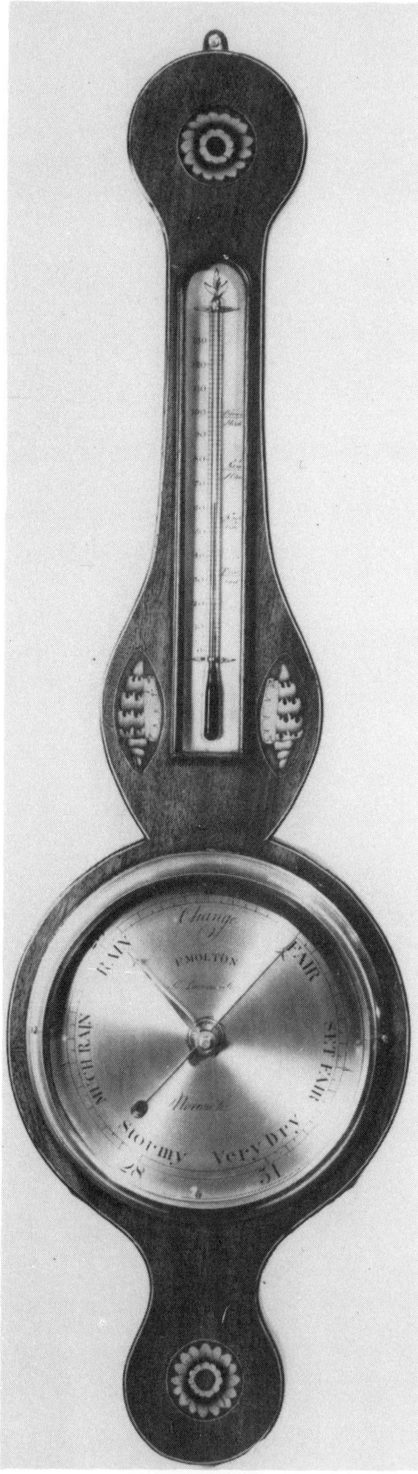
Change
RAIN
FAIR
MUCH RAIN
SET FAIR
Stormy
Very Dry
28
31

150

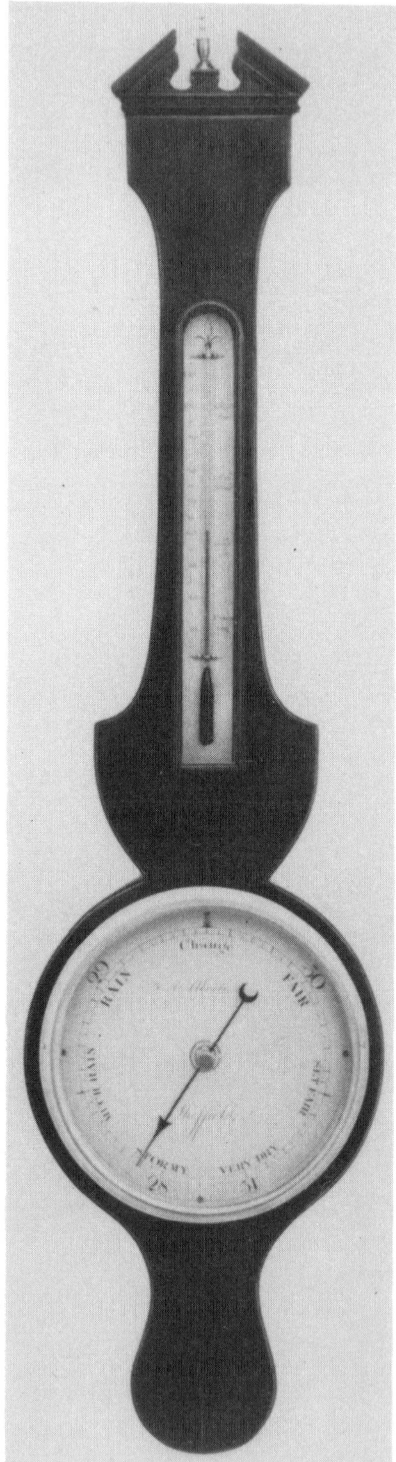
Change
29
RAIN
FAIR
30
MUCH RAIN
SET FAIR
STORMY
VERY DRY
28
31

151

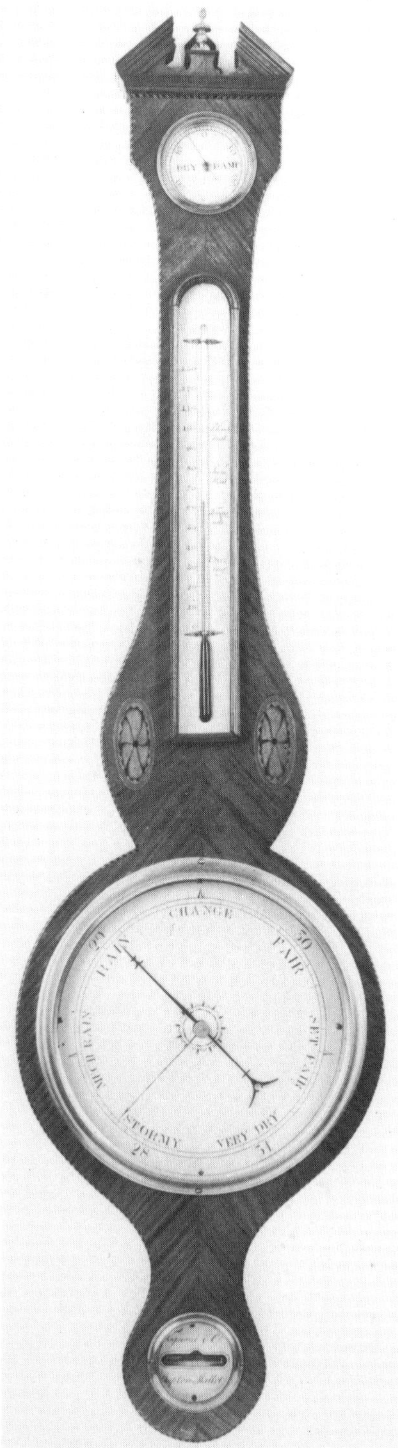
DRY DAMP
CHANGE
29
RAIN
FAIR
30
MUCH RAIN
SET FAIR
STORMY
VERY DRY
28
31

152

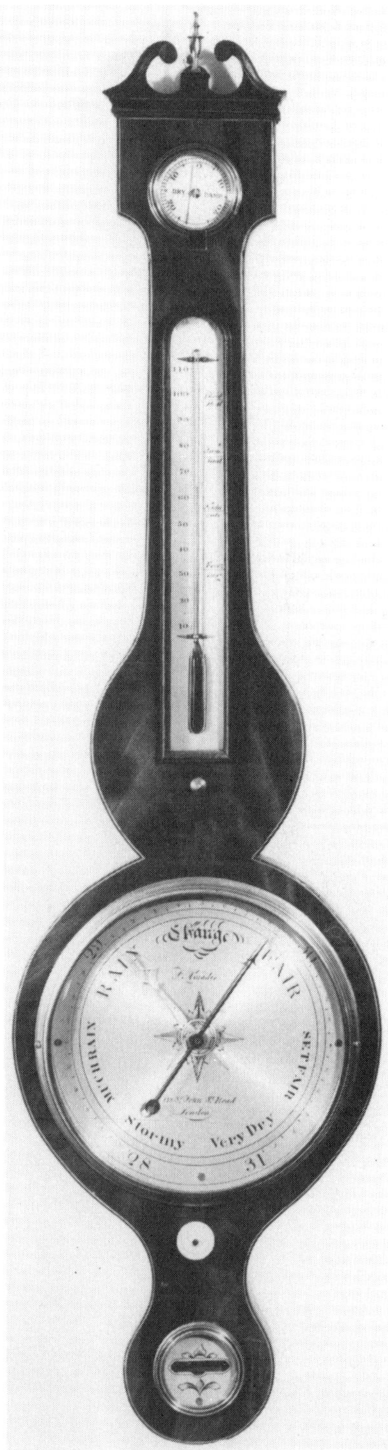
DRY DAMP
Change
RAIN
FAIR
MUCH RAIN
SET FAIR
Stormy
Very Dry
28
31

153

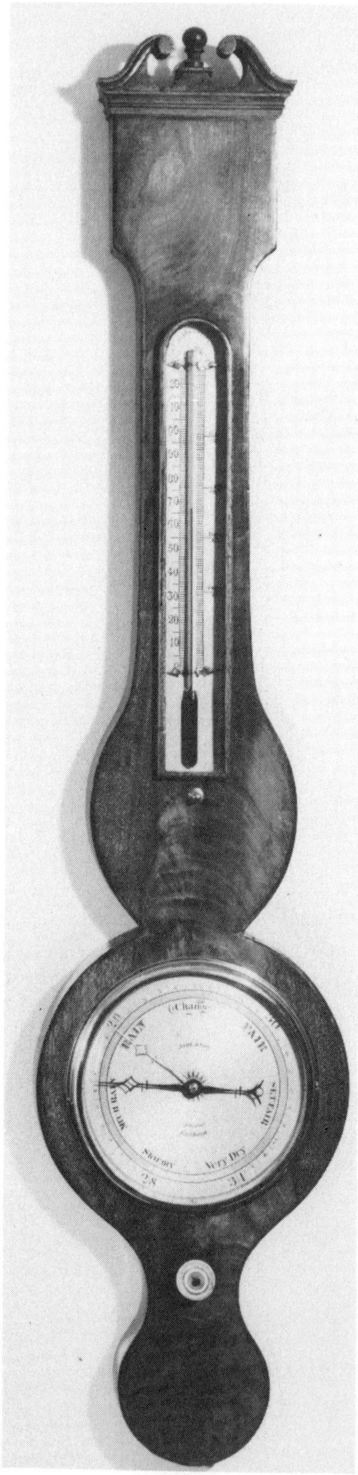

154

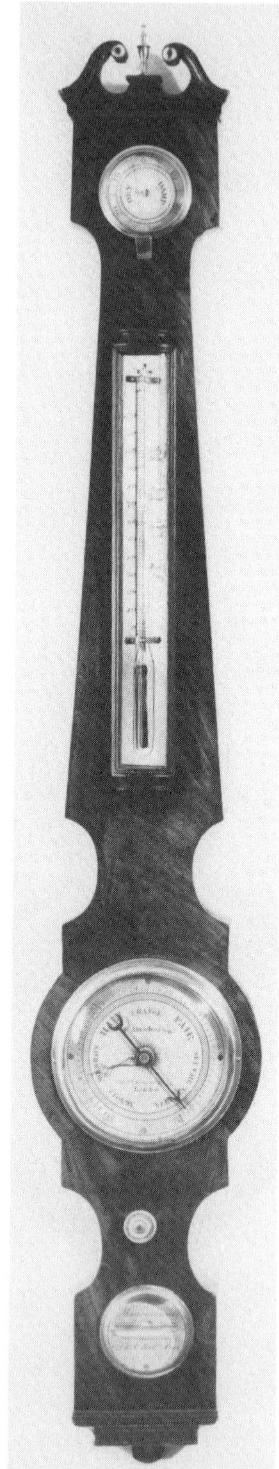

155

Change
RAIN
FAIR
SET FAIR
Stormy
Very Dry
FRENCH

156

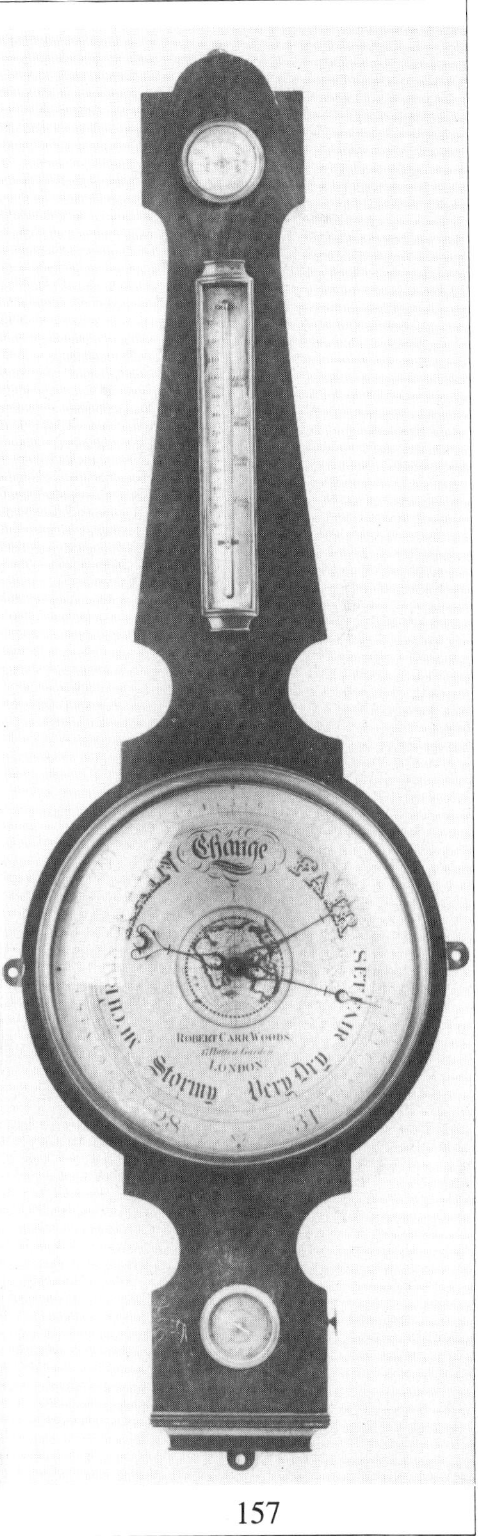
Change
FAIR
SET FAIR
ROBERT CARR WOODS
LONDON
Stormy
Very Dry

157

DIURNAL
REGISTER
29
30
28
31

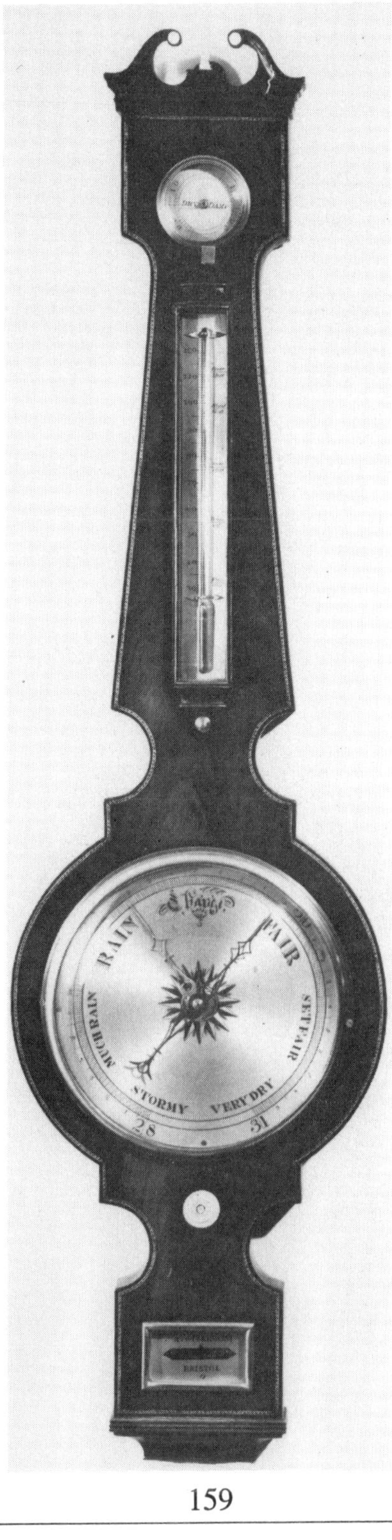
RAIN
FAIR
MUCH RAIN
SET FAIR
STORMY
VERY DRY
28
31
BRISTOL

159

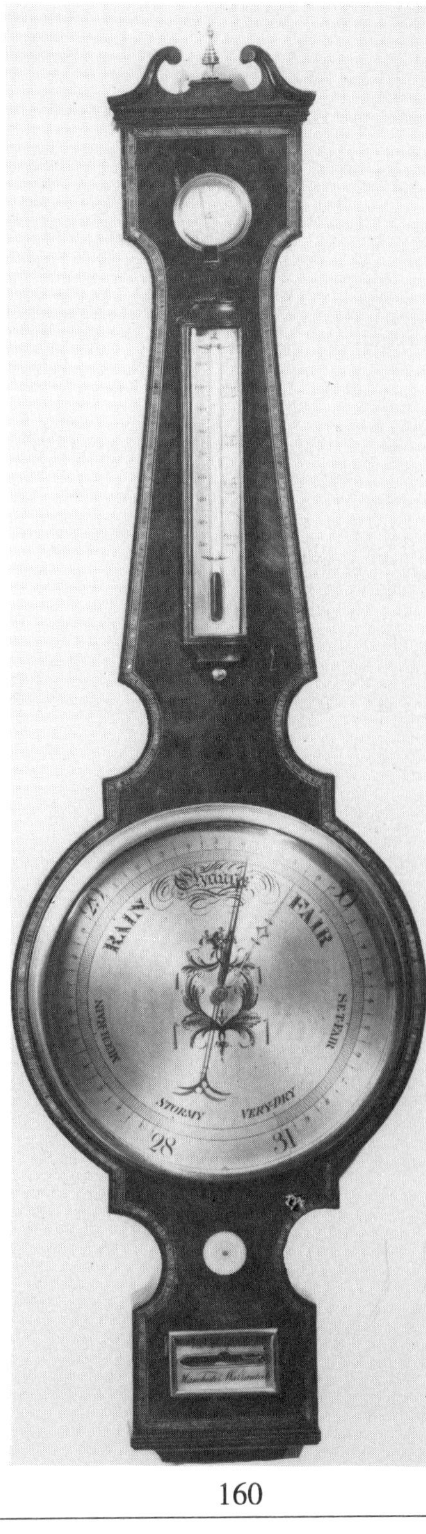
29
30
RAIN
FAIR
MUCH RAIN
SET FAIR
STORMY
VERY-DRY
28
31

160

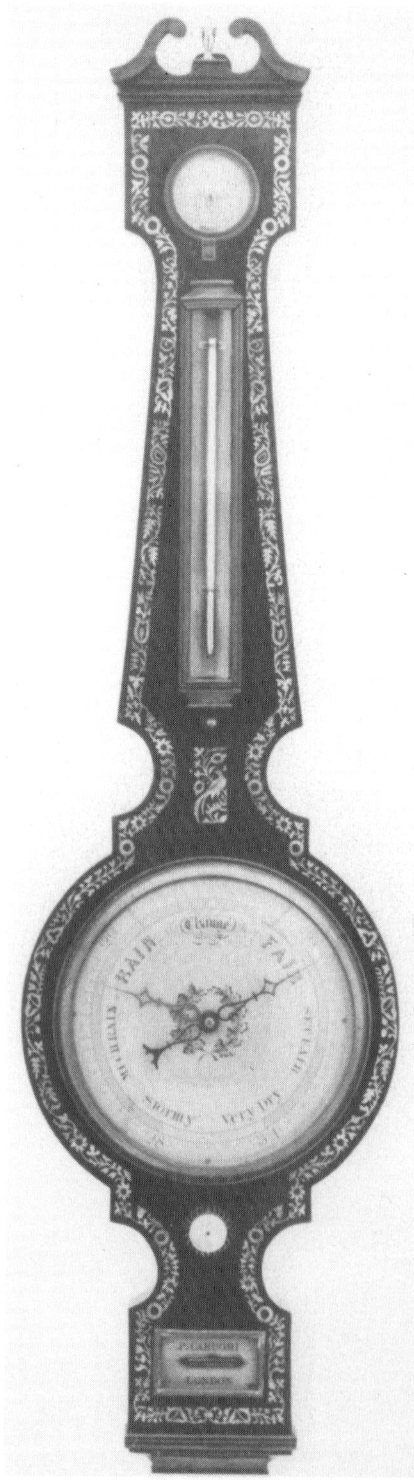
Change
RAIN
FAIR
MUCH RAIN
SET FAIR
STORMY
Very Dry
LONDON

161

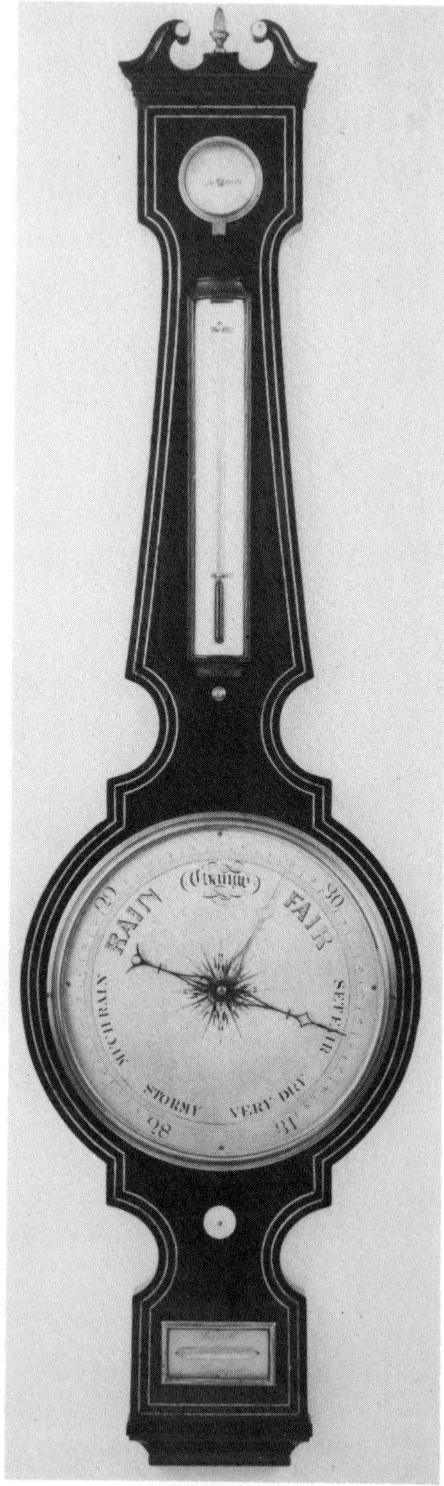
Change
29
RAIN
FAIR
30
MUCH RAIN
SET FAIR
STORMY
VERY DRY
28
31

162

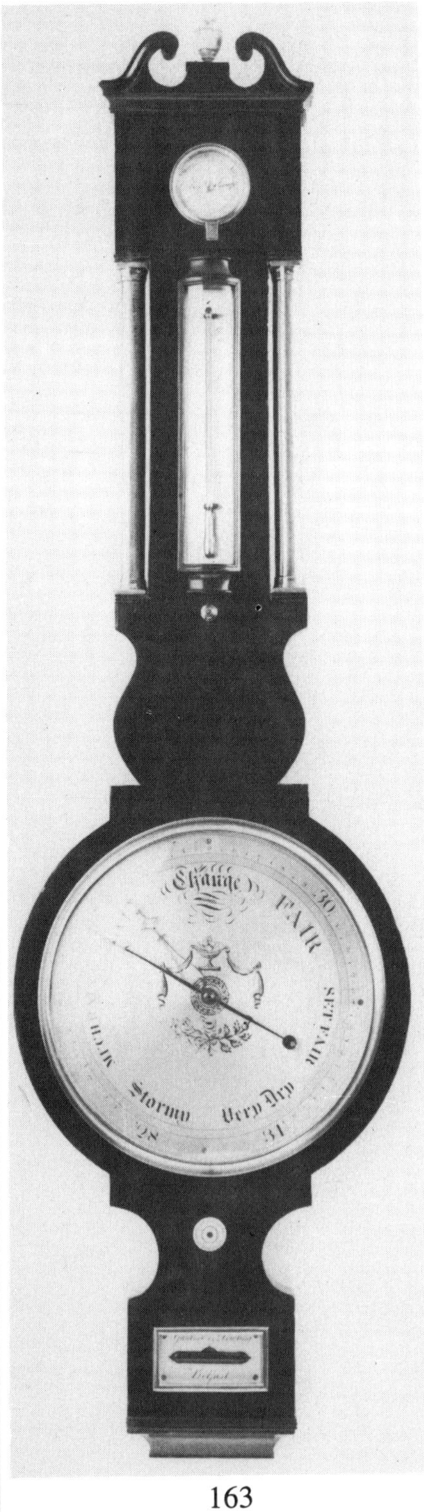
Change
FAIR
30
SET FAIR
Stormy
Very Dry
28
31

163

Change
FAIR
30
SET FAIR
STORMY
VERY DRY
28
31

164

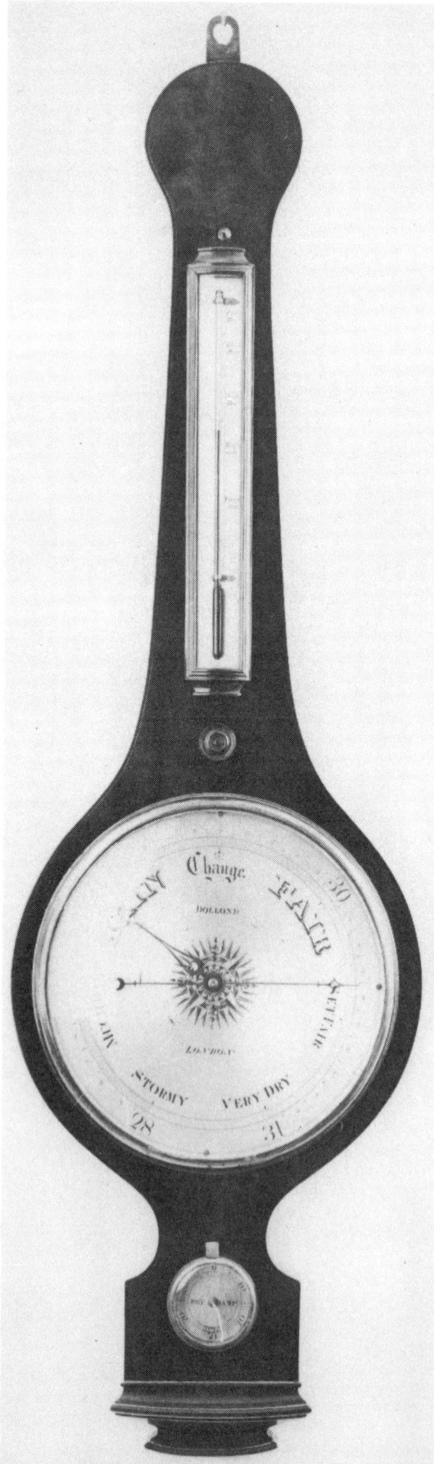
Change
DOLLOND
FAIR
30
SET FAIR
LONDON
STORMY
VERY DRY
28
31

165

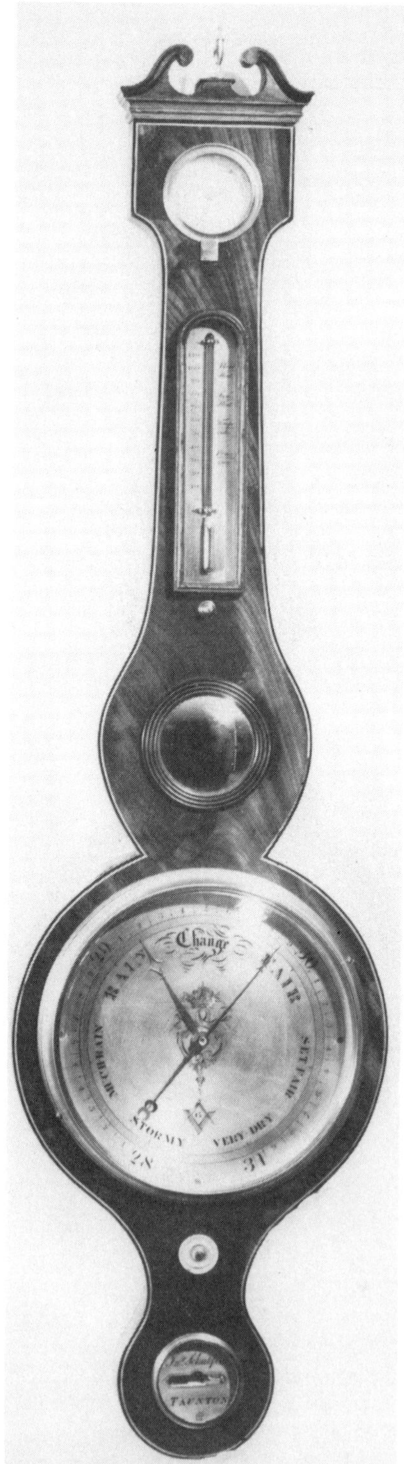
Change
RAIN
FAIR
MUCH RAIN
SET FAIR
STORMY
VERY DRY
28
31
TAUNTON

166

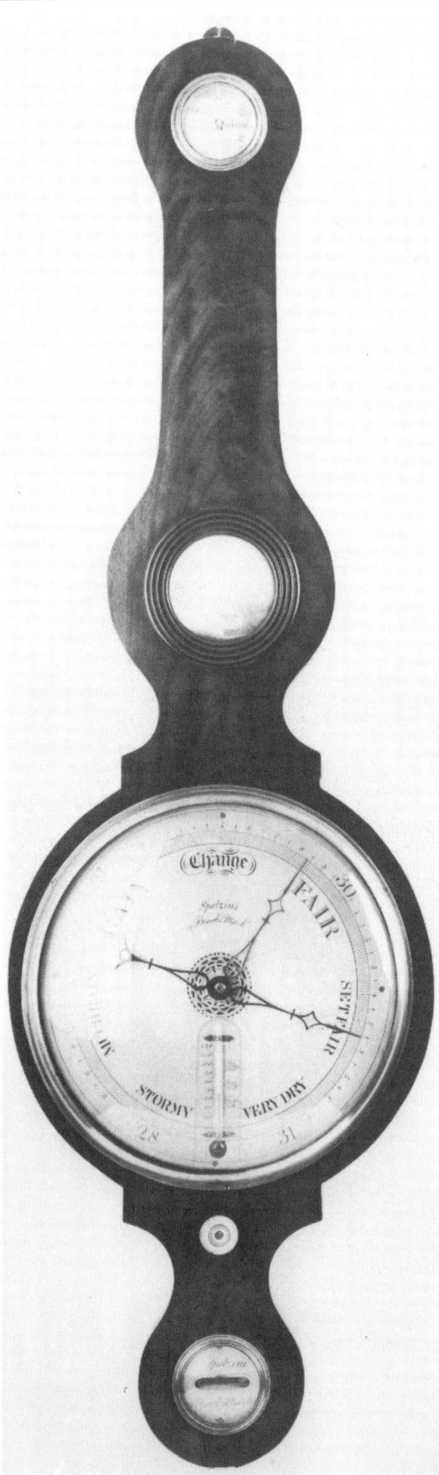
Change
30
FAIR
SET FAIR
STORMY
VERY DRY
28
31

167

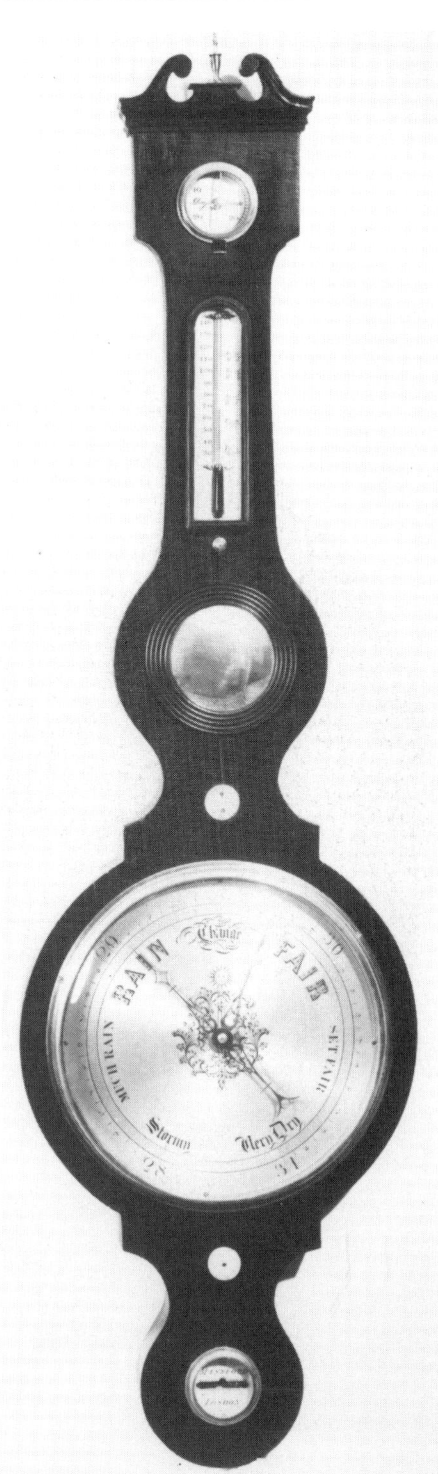
Change
29
RAIN
FAIR
30
MUCH RAIN
SET FAIR
Stormy
Very Dry
28
31

168

Change
RAIN
FAIR
29
30
MUCH RAIN
SET FAIR
Stormy
Very Dry
28
31

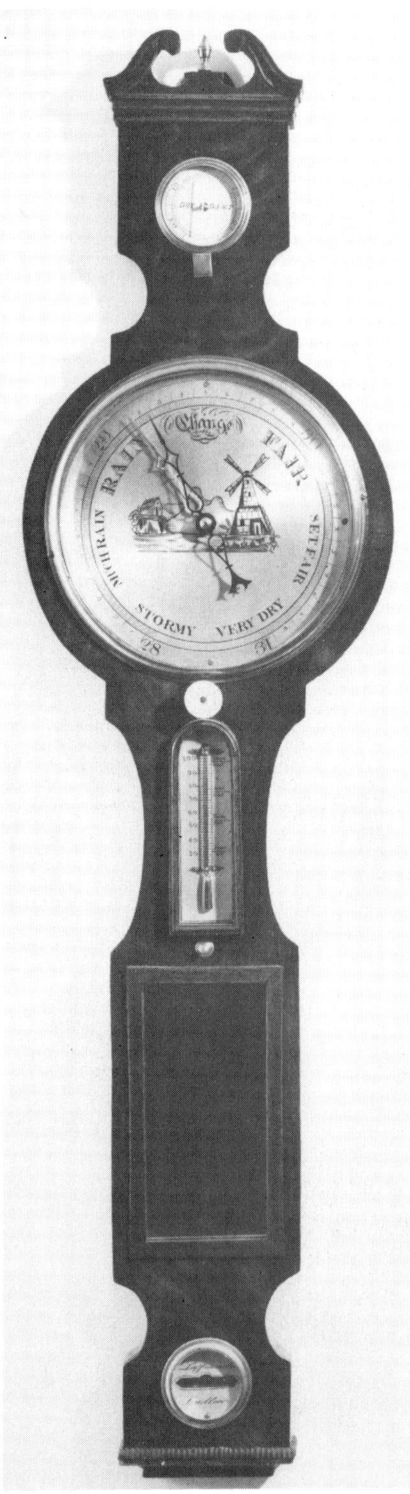
Change
RAIN
FAIR
MUCH RAIN
SET FAIR
STORMY
VERY DRY
28
31

170

Change
29
30
RAIN
FAIR
MUCH RAIN
SET FAIR
STORMY
VERY DRY
28
31

171

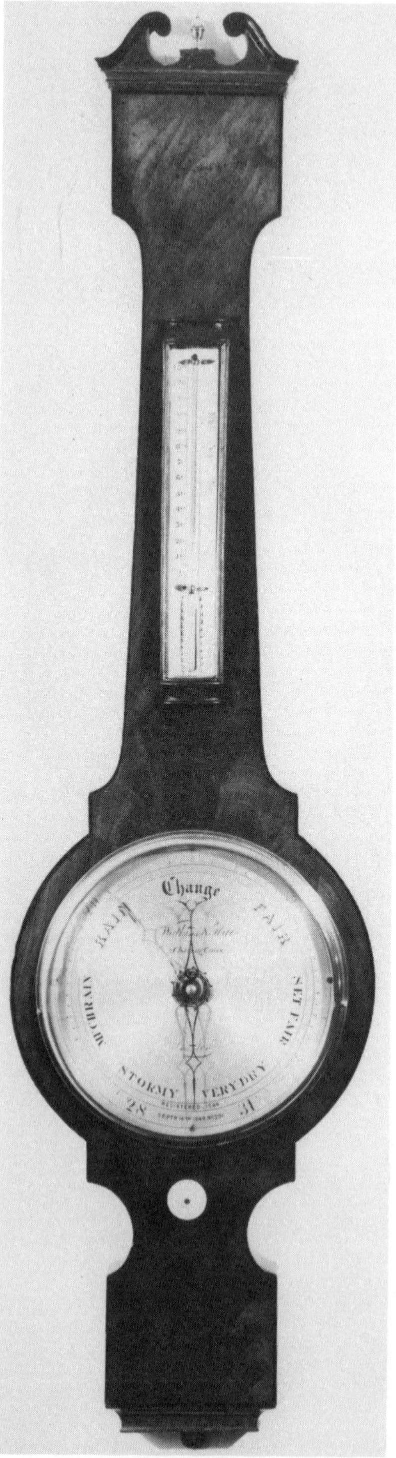
Change
RAIN
FAIR
SET FAIR
STORMY
VERY DRY
28
31

172

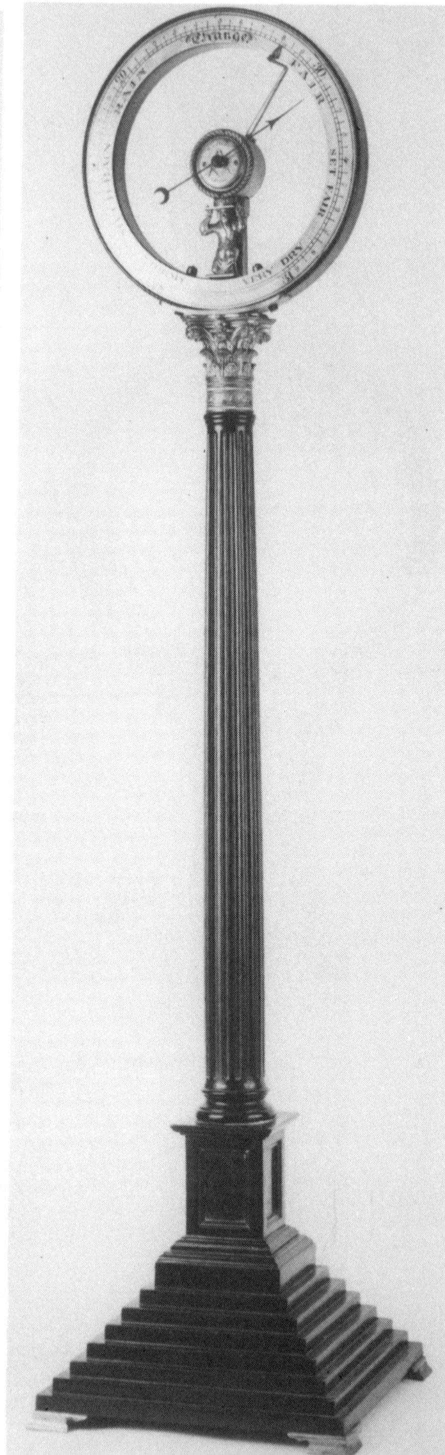
Change
RAIN
FAIR
SET FAIR
VERY DRY

173

Change
FAIR
SET FAIR
VERY DRY
STORMY
MUCH RAIN
RAIN
28
29
30
31
LONDON

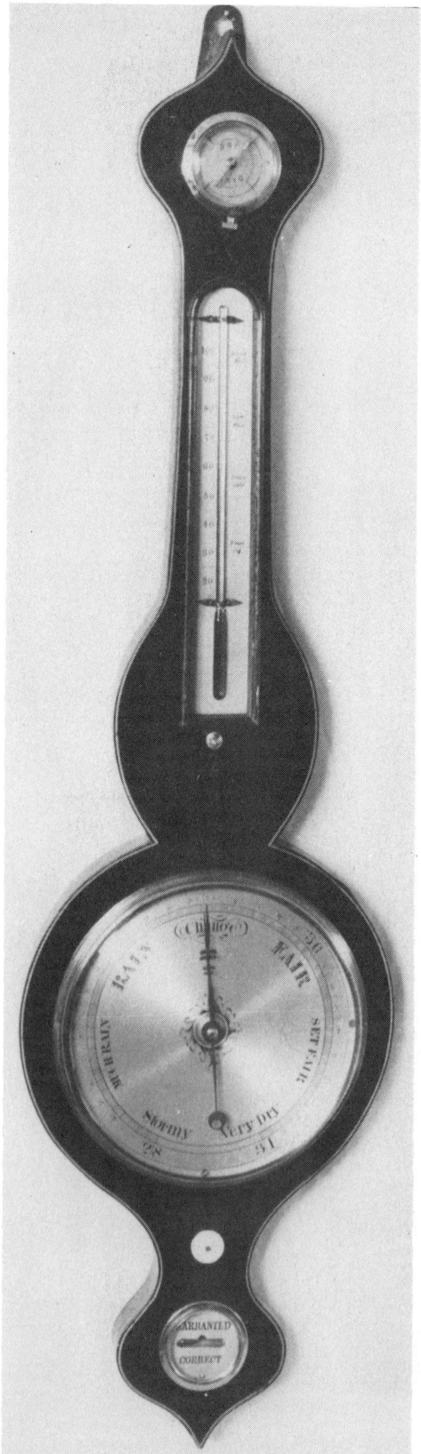
Change
RAIN
FAIR
SET FAIR
Stormy
Very Dry
28
31
CORRECT

175

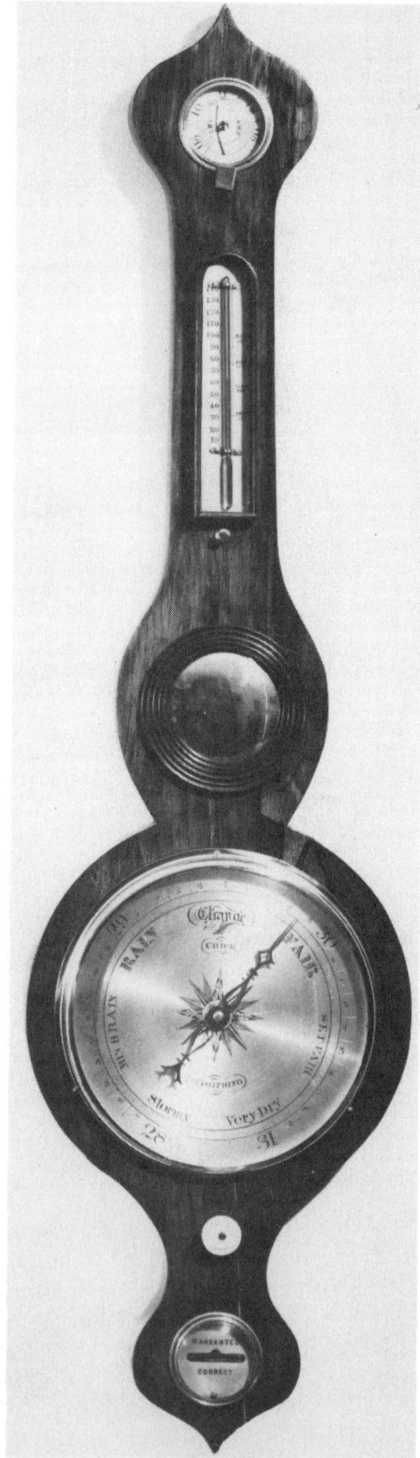
Change
RAIN
FAIR
30
SET FAIR
Stormy
Very Dry
28
31
WARRANTED
CORRECT

176

29
RAIN
Change
FAIR
30
SET FAIR
Stormy
Very Dry
28
31
HEREFORD

177

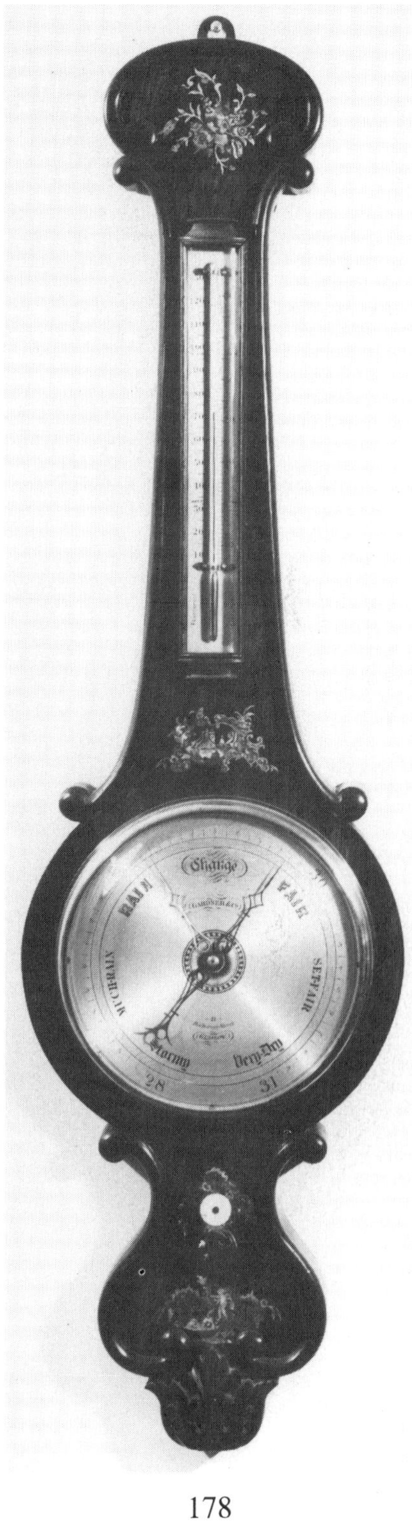
Change
RAIN
FAIR
SET FAIR
Stormy
Very Dry
28
31

178

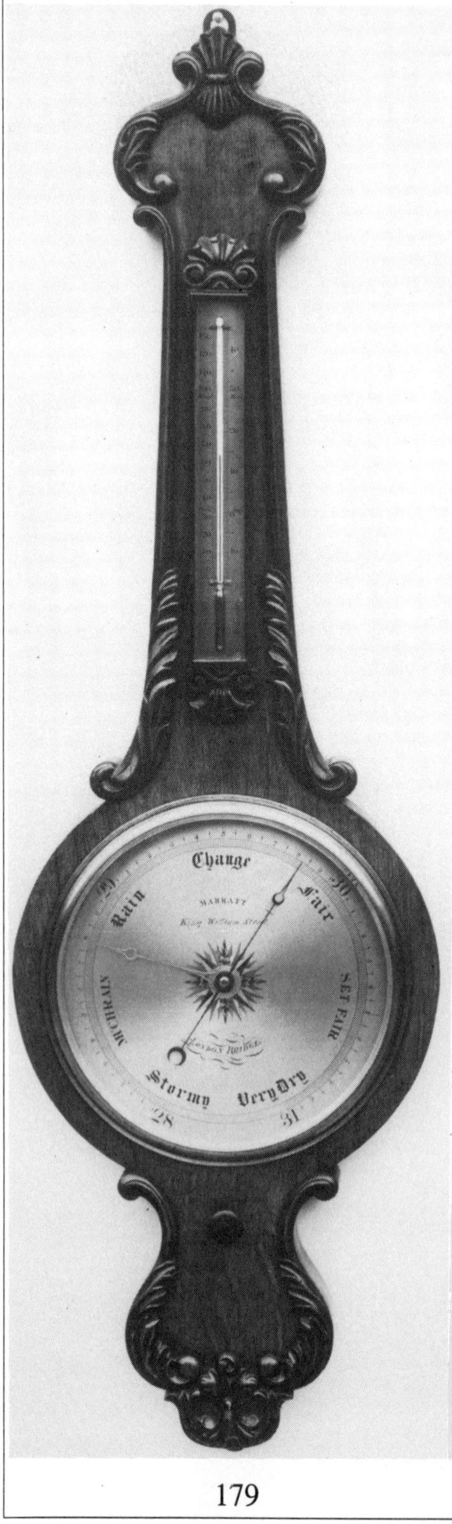

Change
Rain
Fair
MARRATT
M'CH RAIN
SET FAIR
Stormy
Very Dry
28
31

179

Change
29
30
RAIN
FAIR
M'CH RAIN
SET FAIR
Stormy
Very Dry

180

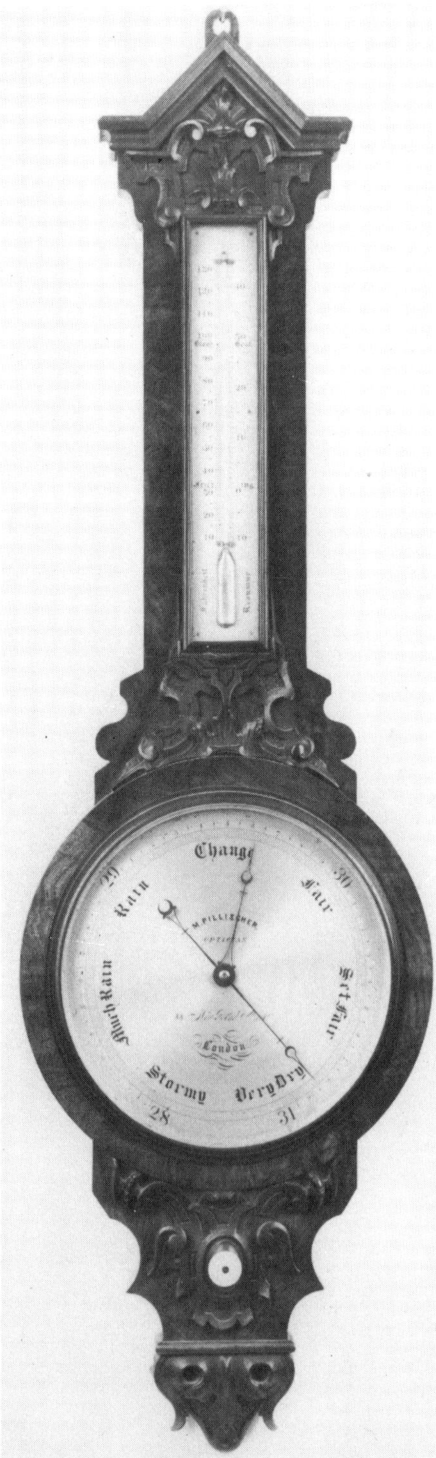
Change
29
Rain
Fair
30
M. PILLISCHER
OPTICIAN
Much Rain
Set Fair
London
Stormy
Very Dry
28
31

181

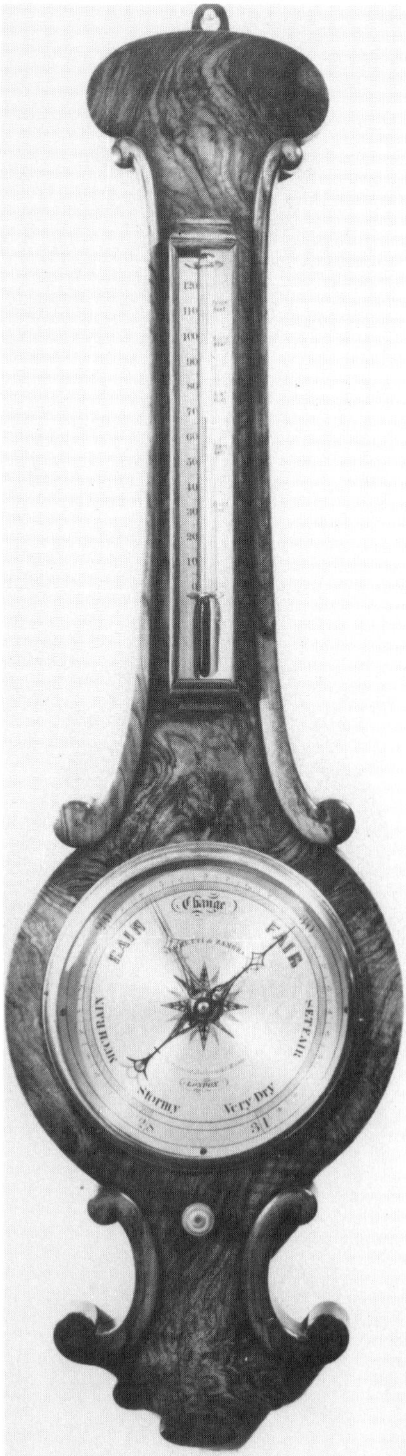
Change
RAIN
FAIR
MUCH RAIN
SET FAIR
Stormy
Very Dry
28
31

182

Change
RAIN
FAIR
SET FAIR
Stormy
Very Dry
29
30
28
31
183

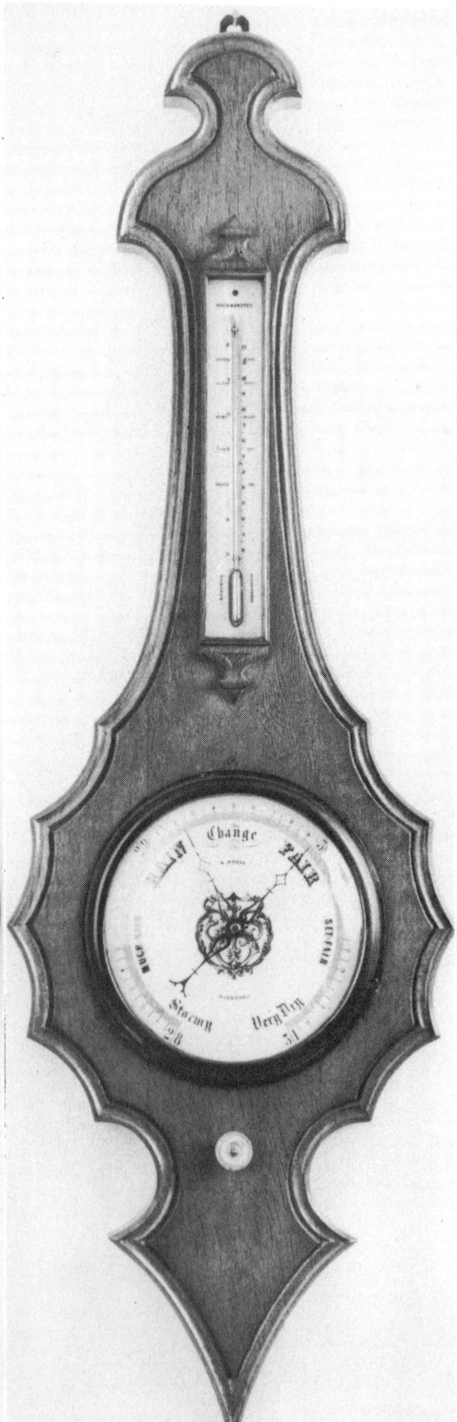
Change
RAIN
FAIR
SET FAIR
Stormy
Very Dry
28
31
184

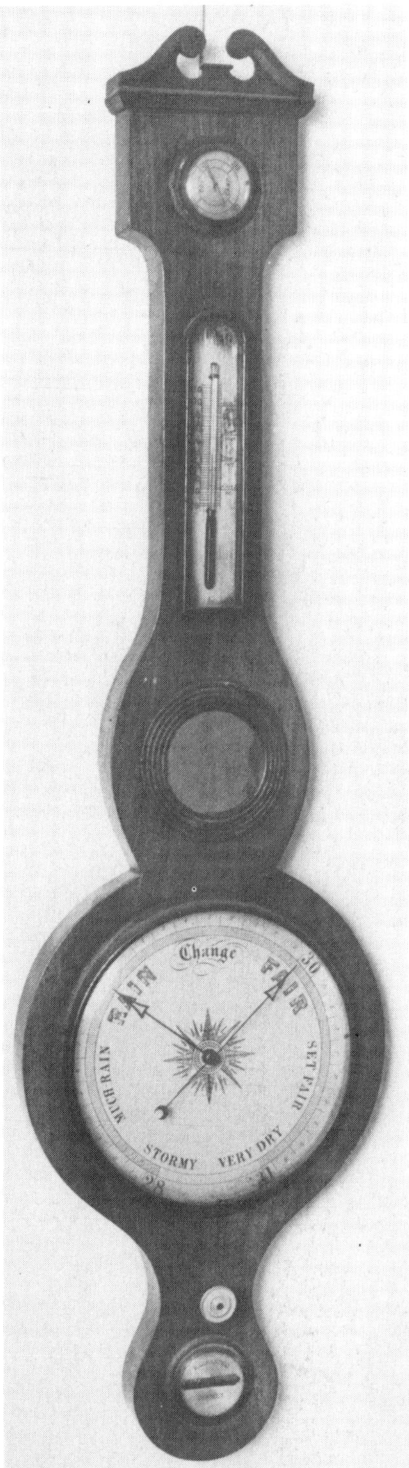
Change
RAIN
FAIR
30
MUCH RAIN
SET FAIR
STORMY
VERY DRY
28
31

185

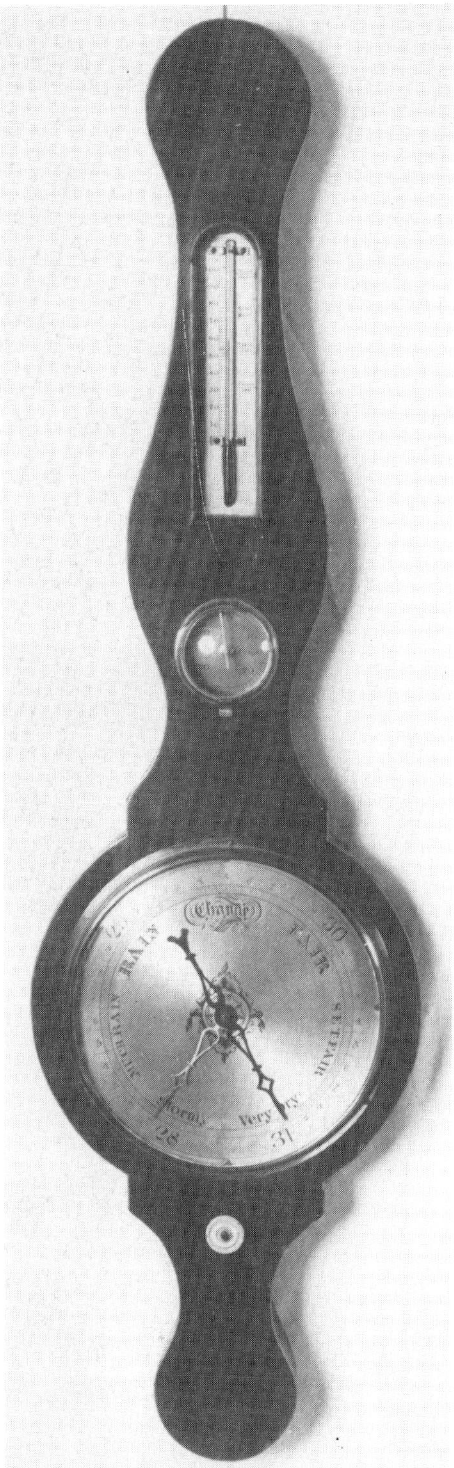
Change
RAIN
FAIR
30
MUCH RAIN
SET FAIR
28
31

186

187

CHANGE
RAIN
FAIR
MUCH RAIN
SET FAIR
STORMY
VERY DRY
30
28
31
METALLIC
BAROMETER

188

CHANGE
RAIN
FAIR
MUCH RAIN
SET FAIR
STORMY
VERY DRY
29
30
28
31
METALLIC
BAROMETER

189

190

191

192

193

28
29
30
31
STORMY
RAIN
CHANGE
FAIR
VERY DRY
FAHRENHEIT
CENTIGRADE

194

195

STORMY
RAIN
CHANGE
FAIR
VERY DRY

196

STORMY
CHANGE
VERY DRY
COMPENSATED

197

Negretti & Zambra
LONDON
5680

198

STORMY. Rain CHANGE Fair. VERY DRY.
Aneroid Barometer

199

RAIN CHANGE FAIR
Compensated

200

201

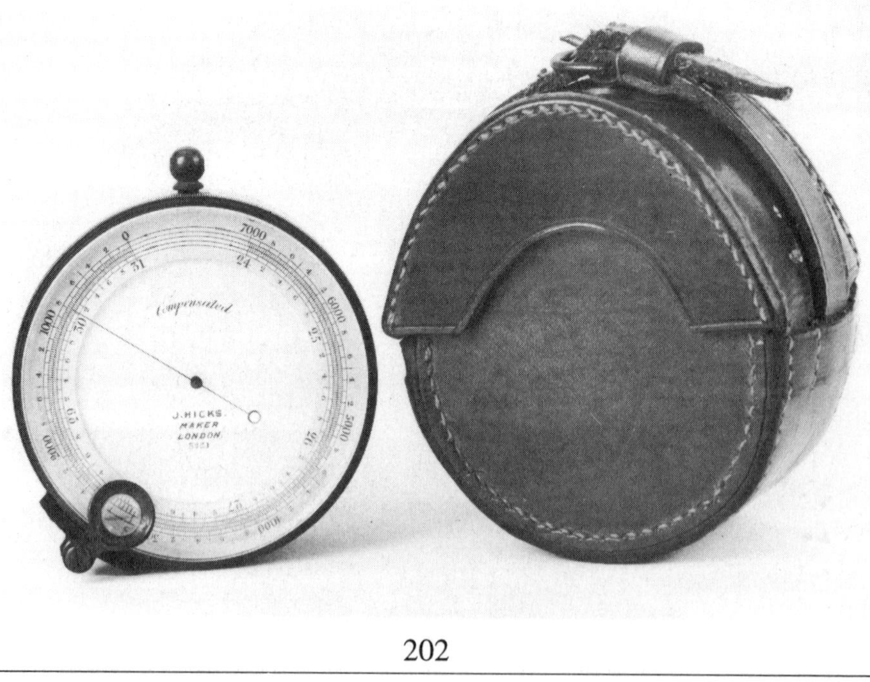

202

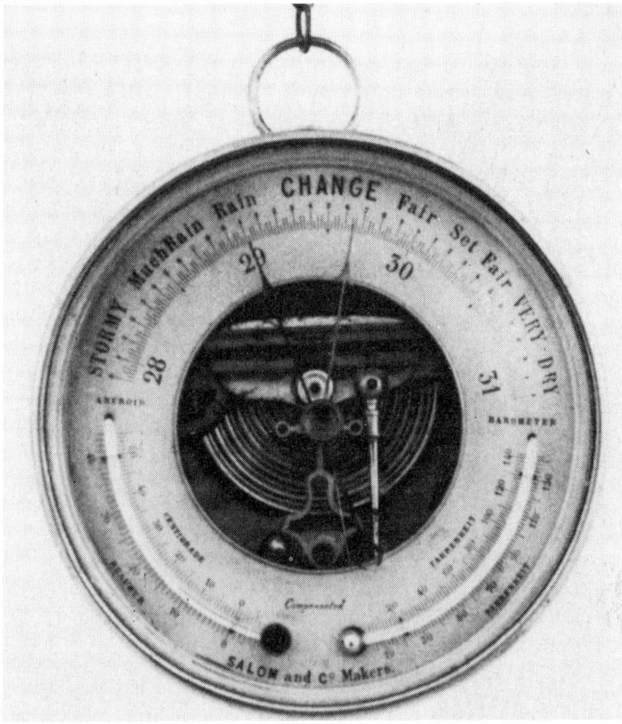
CHANGE
Rain
Fair
Set Fair
VERY DRY
STORMY
28
29
30
31
ANEROID
BAROMETER
Compensated
SALOM and Co Makers.

203

CHANGE
RAIN
FAIR
STORMY
VERY DRY
WET
MORE WIND
DRY
LESS WIND
28
29
30
31
FALL
RISE
FAHRENHEIT

204

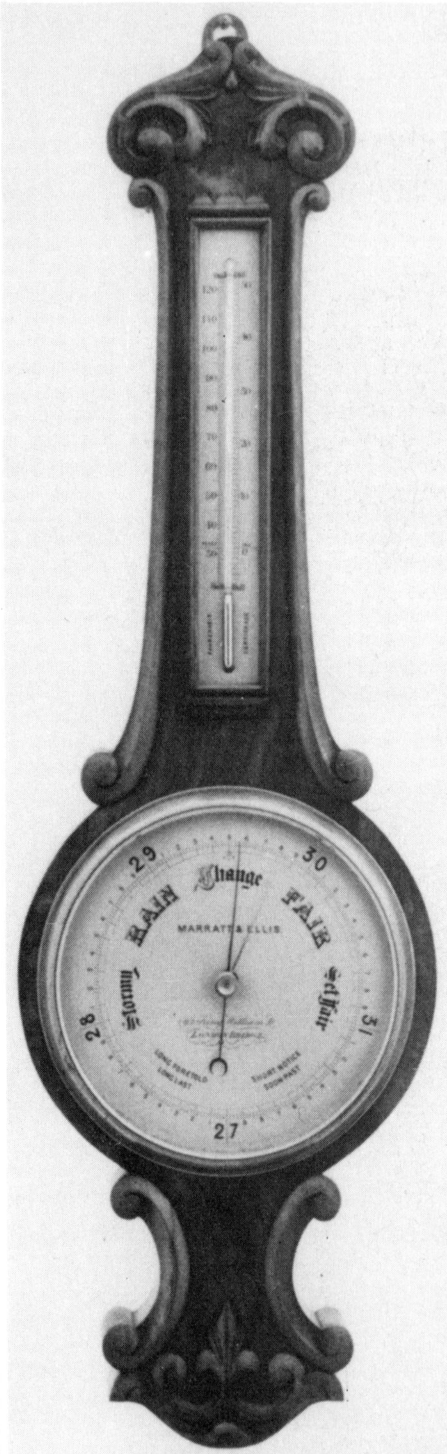
29
30
Change
RAIN
FAIR
MARRATT & ELLIS
Stormy
Set Fair
28
31
LONG FORETOLD
LONG LAST
SHORT NOTICE
SOON PAST
27

205

STORMY
MUCH RAIN
RAIN
CHANGE
FAIR
SET FAIR
VERY DRY
WET
DRY

206

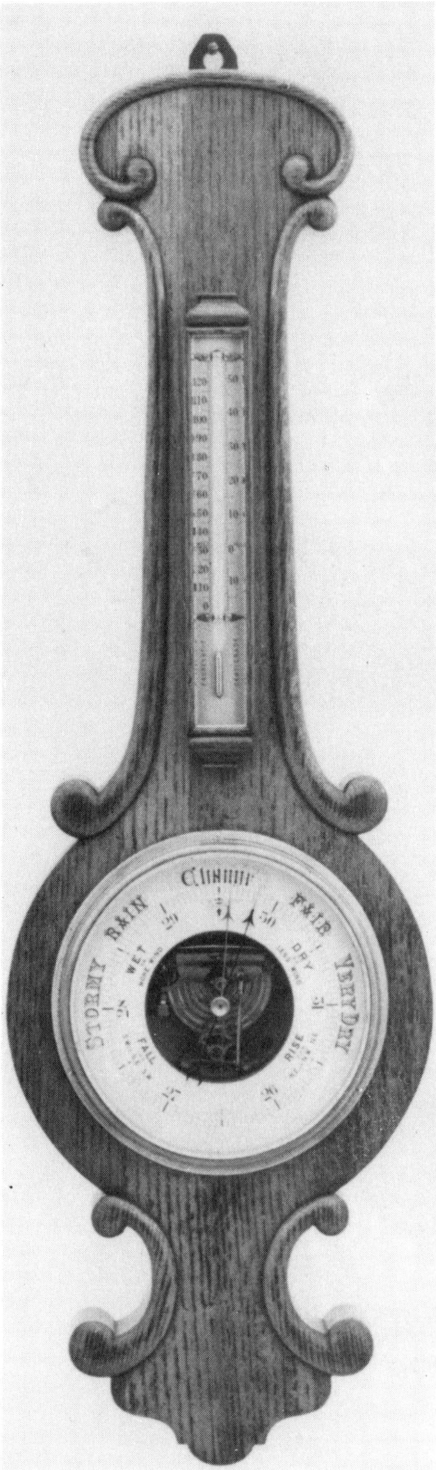
STORMY
RAIN
CHANGE
FAIR
VERY DRY
WET
DRY
FALL
RISE

207

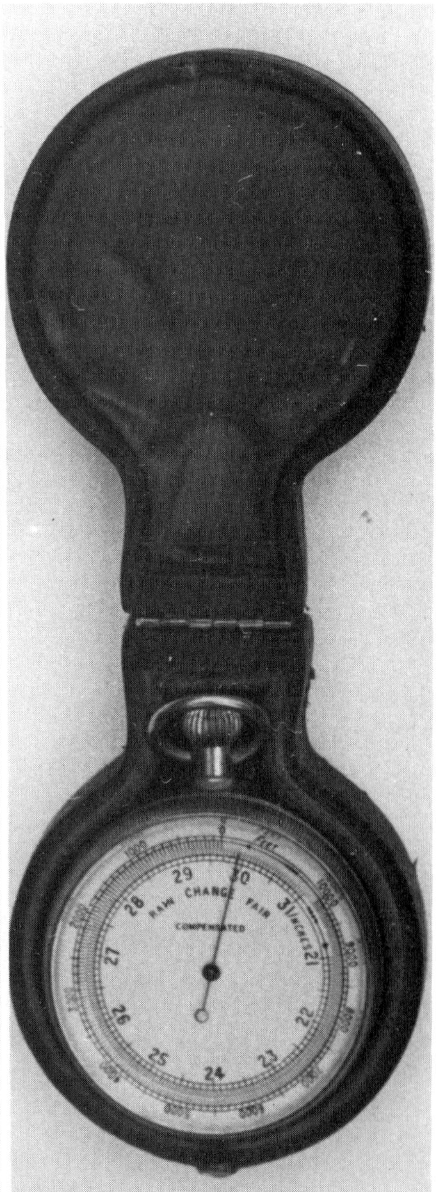
RAIN
CHANGE
FAIR
COMPENSATED
INCHES

208

CHANGE
Rain
Fair

209

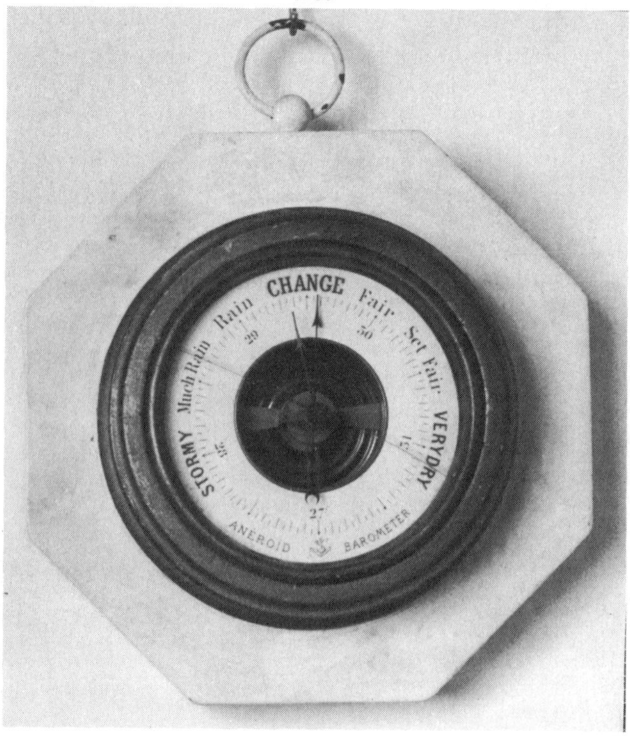
CHANGE
Rain
Fair
Much Rain
Set Fair
STORMY
VERYDRY
29
30
27
ANEROID
BAROMETER

210

211

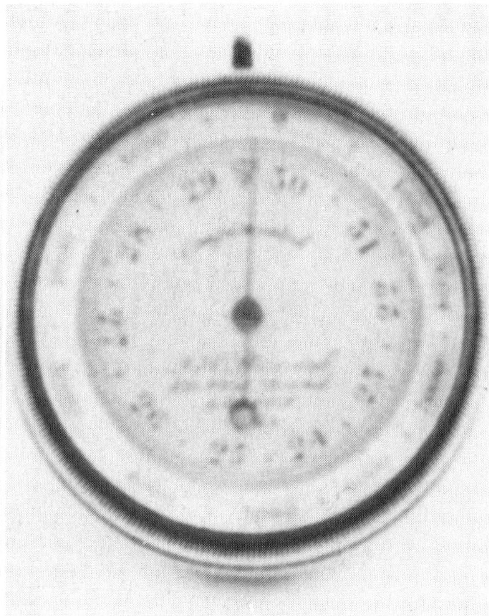

212

213

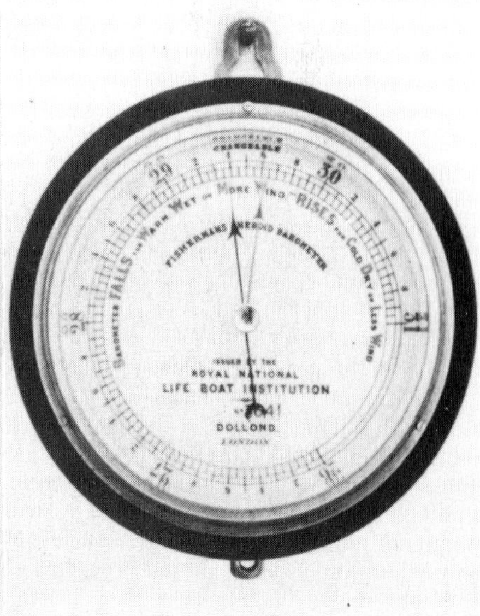

214

CHANGEABLE
29
30
31
FALLS FOR WARM, WET OR MORE WIND
RISES FOR COLD, DRY OR LESS WIND
MARINE ANEROID BAROMETER
MADE BY
DOLLOND
LONDON.
THE
SHIPWRECKED FISHERMEN & MARINERS'
26. SUFFOLK STREET.
PALL MALL. EAST. S.W.
ROYAL BENEVOLENT SOCIETY

215

UNSETTLED
STORMY
FAIR WEATHER
28
29
30
31
BAROMETER FALLS
FOR WARM, WET OR MORE WIND
BAROMETER RISES
FOR COLD, DRY OR LESS WIND
THE
"UNIVERSAL BAROMETER"
FOR THE USE OF
MARINERS AGRICULTURISTS HORTICULTURISTS &c

216

ATMOS
BAROMETER
STORMY
RAIN
CHANGE
FAIR
VERY-DRY
28
29
30
31
FAHRENHEIT

217

STORMY
RAIN
CHANGE
FAIR

218

29
30
RAIN
CHANGE
FAIR
31
28
27
23
24
25
26
LIVERPOOL
4000
5000

219 220

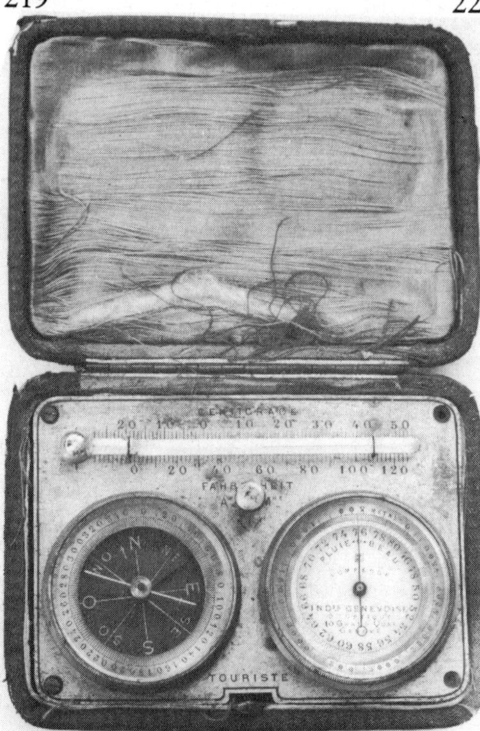
20 10 0 10 20 30 40 50
0 20 40 60 80 100 120
N
E
S
PLUIE
BEAU
INDU GENEVOISE
TOURISTE

221

222 223 224

225 226 227

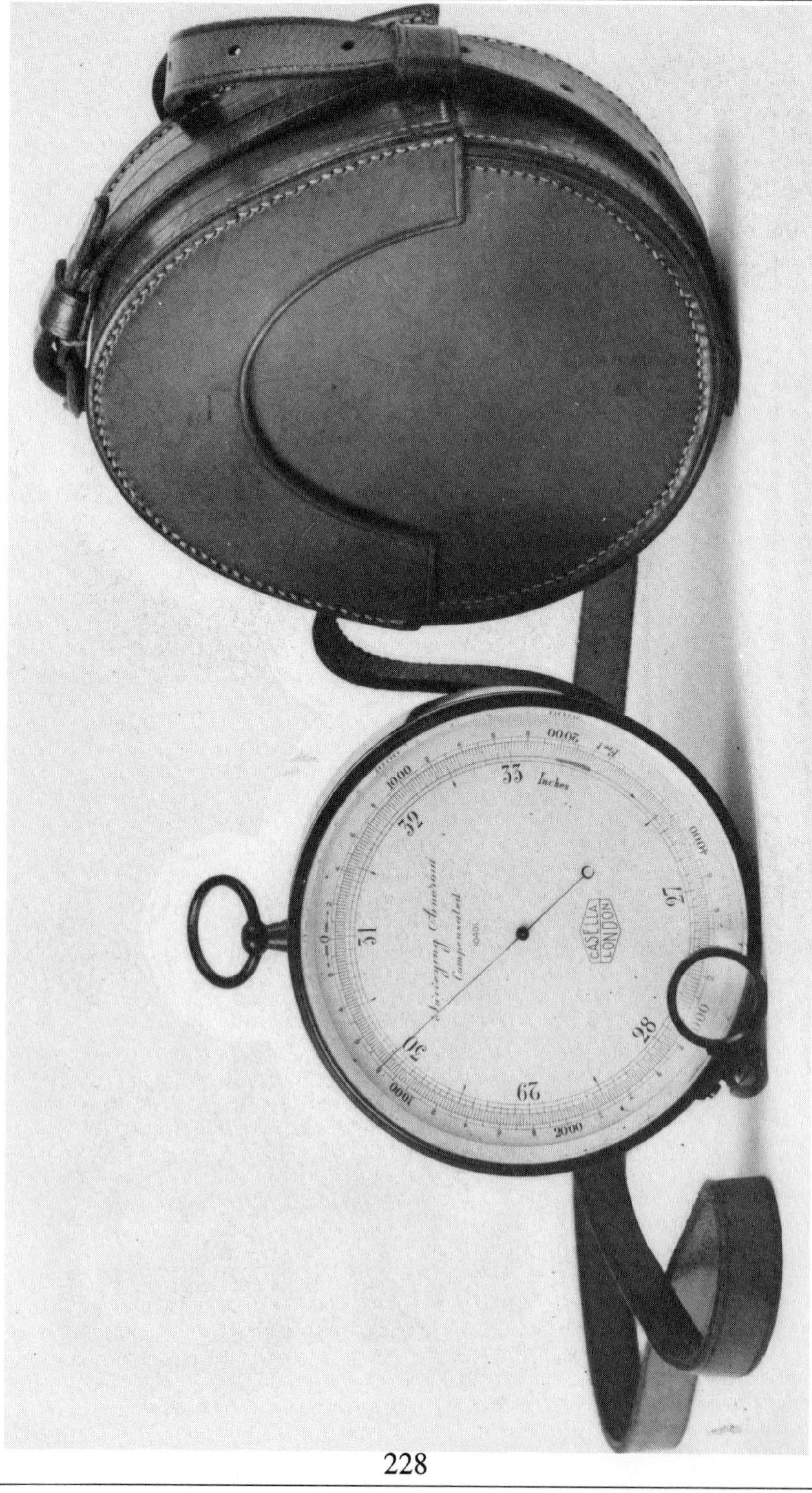
CASELLA
LONDON
Inches

Change
Rain
Fair
Stormy
Very Dry
Wet
More Wind
Dry
Less Wind
Fall
Rise
29
30
28
31
27
26
Aneroid Barometer

229

Change
RAIN
FAIR
STORMY
VERY DRY
London

230

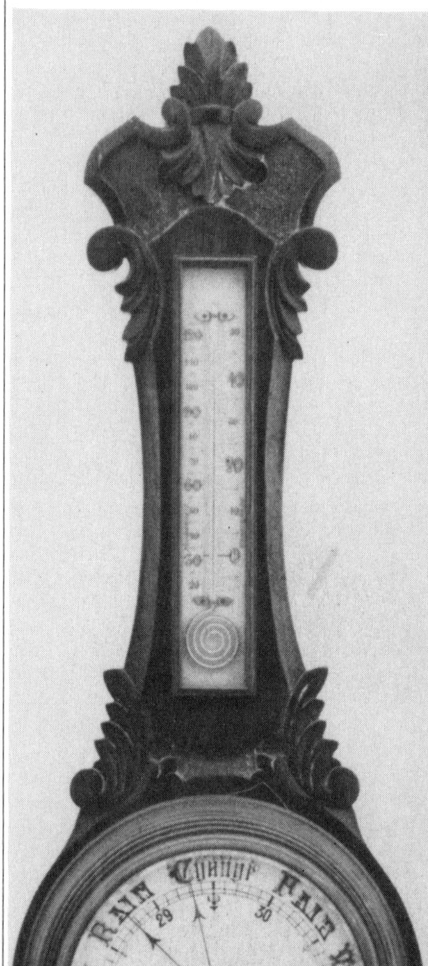
Stormy Rain Change Fair Very Dry
28 29 30 31 27 26

231

Stormy Rain Change Fair Very Dry
28 29 30 31 27 26

232

233

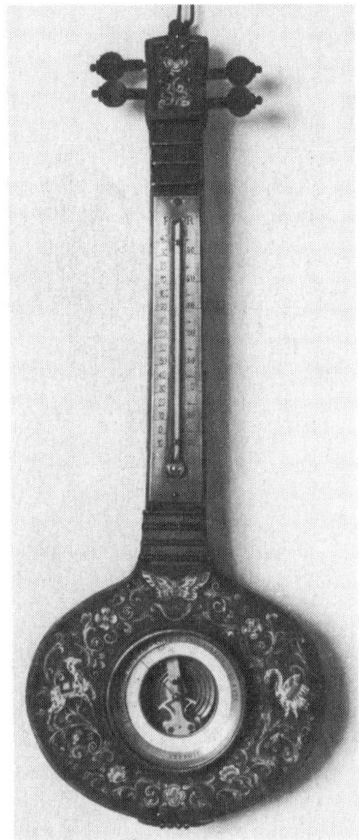

234

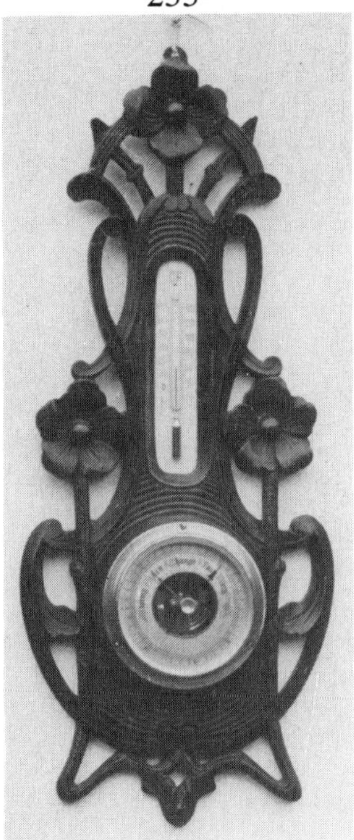

235

236

CHANGE
RAIN
FAIR
SET FAIR
Aneroid Barometer

Stormy
Change
Fair

0 FEET
Compensated
Inches
JOHN WARDALE & Co
LONDON
No 633
1918

238

239

COMPENSATED
1000 Feet

240

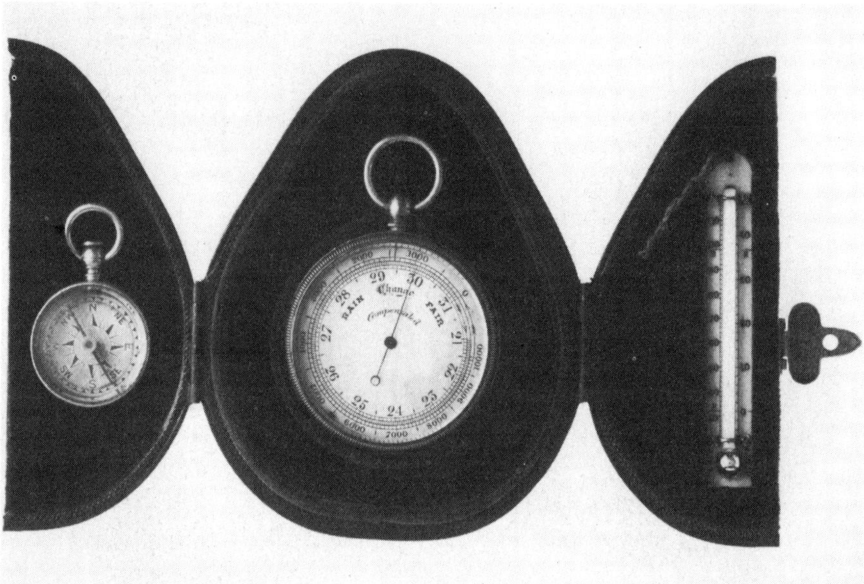

241

242

243

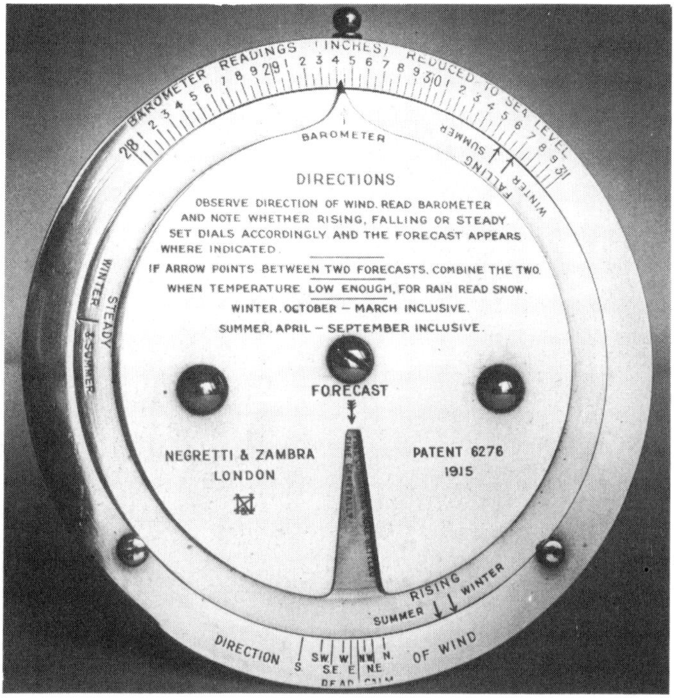
BAROMETER READINGS (INCHES) REDUCED TO SEA LEVEL
BAROMETER
SUMMER
WINTER
FALLING
DIRECTIONS
OBSERVE DIRECTION OF WIND. READ BAROMETER AND NOTE WHETHER RISING, FALLING OR STEADY. SET DIALS ACCORDINGLY AND THE FORECAST APPEARS WHERE INDICATED.
IF ARROW POINTS BETWEEN TWO FORECASTS, COMBINE THE TWO.
WHEN TEMPERATURE LOW ENOUGH, FOR RAIN READ SNOW.
WINTER. OCTOBER – MARCH INCLUSIVE.
SUMMER. APRIL – SEPTEMBER INCLUSIVE.
WINTER
STEADY
SUMMER
FORECAST
NEGRETTI & ZAMBRA
LONDON
PATENT 6276
1915
RISING
SUMMER
WINTER
DIRECTION
OF WIND

244

DIRECTION
OF WIND
SET TO
BAROMETER
READING
FALLING
RISING
STEADY
BAROMETER AT SEA LEVEL
NEGRETTI & ZAMBRA, LONDON

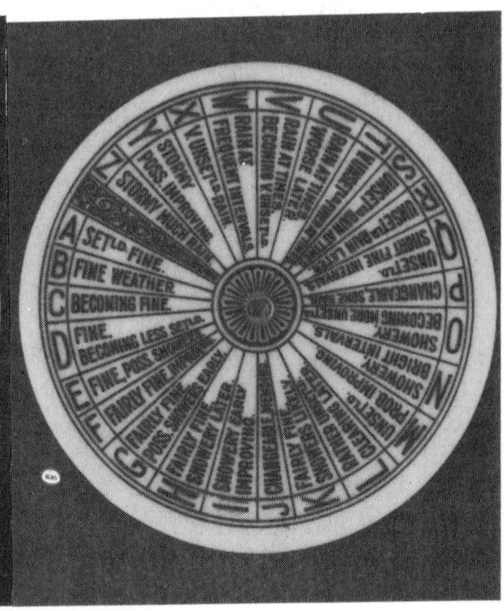
SET'LD FINE.
FINE WEATHER
BECOMING FINE.
FINE. BECOMING LESS SET'LD
FINE. POSS. SHOWERS
FAIRLY FINE
STORMY
STORMY, MUCH RAIN
SHOWERY, BRIGHT INTERVALS

245

246

SEA LEVEL
29
30
28
31
NEGRETTI & ZAMBRA
STEADY
RISE
DIRECTION OF WIND
DEAD CALM

247

248

Fail
Rise
29
30

249

250

251

252

253

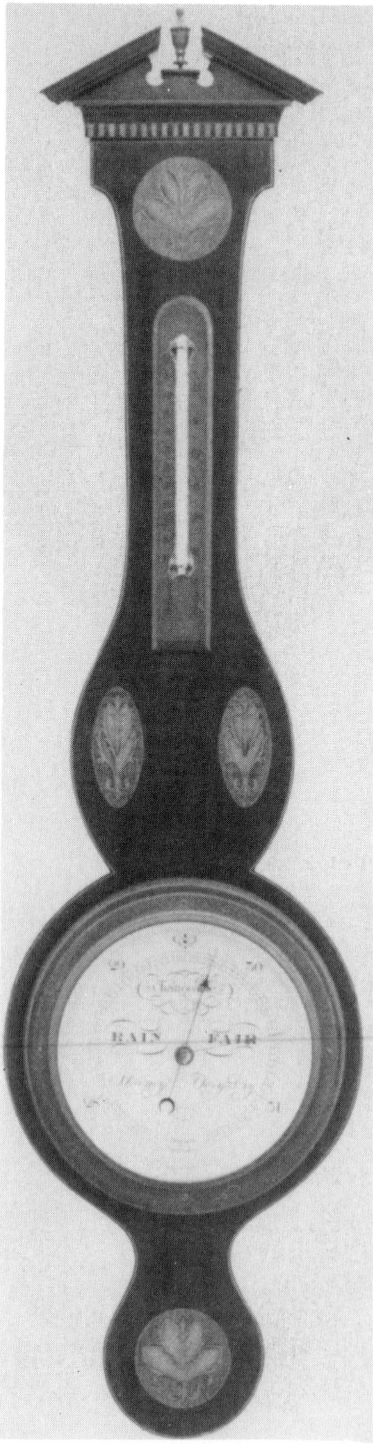

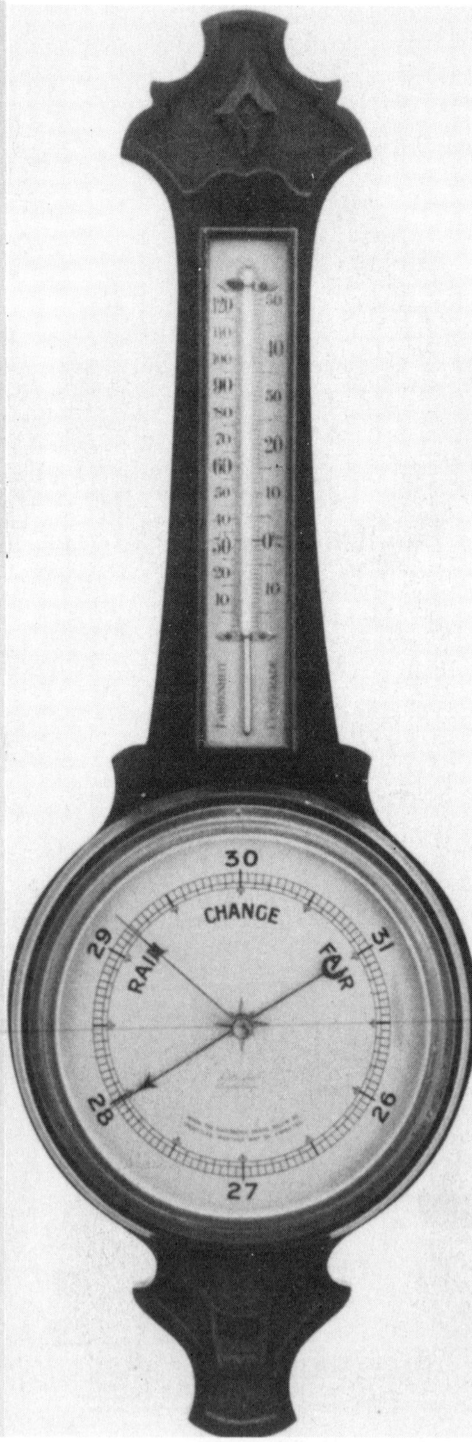

254

255

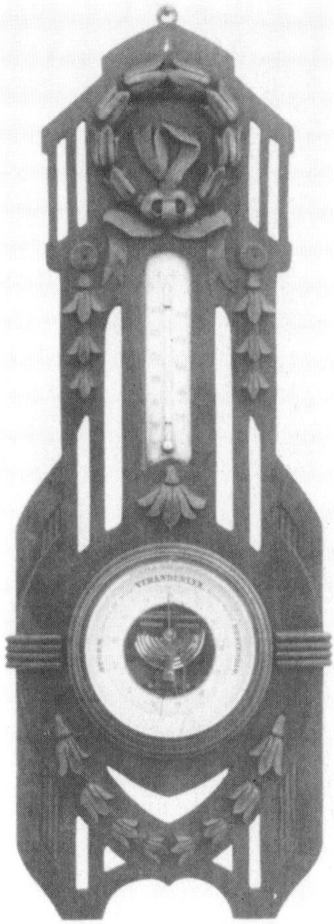

256

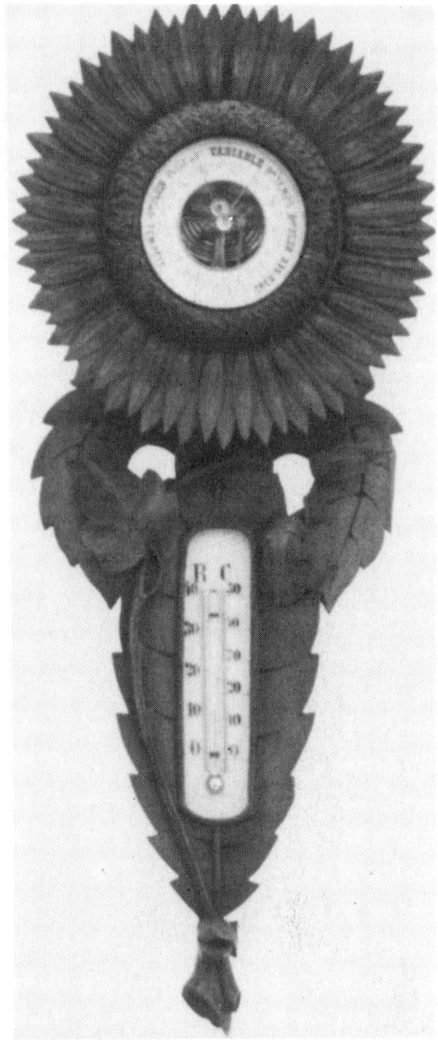

257

258

259

260

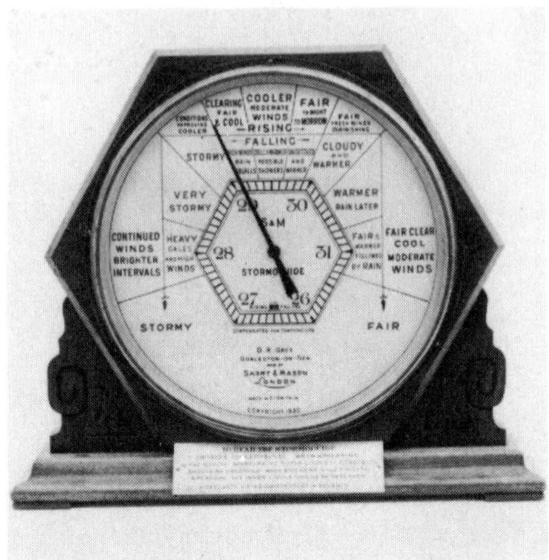

261

262

263

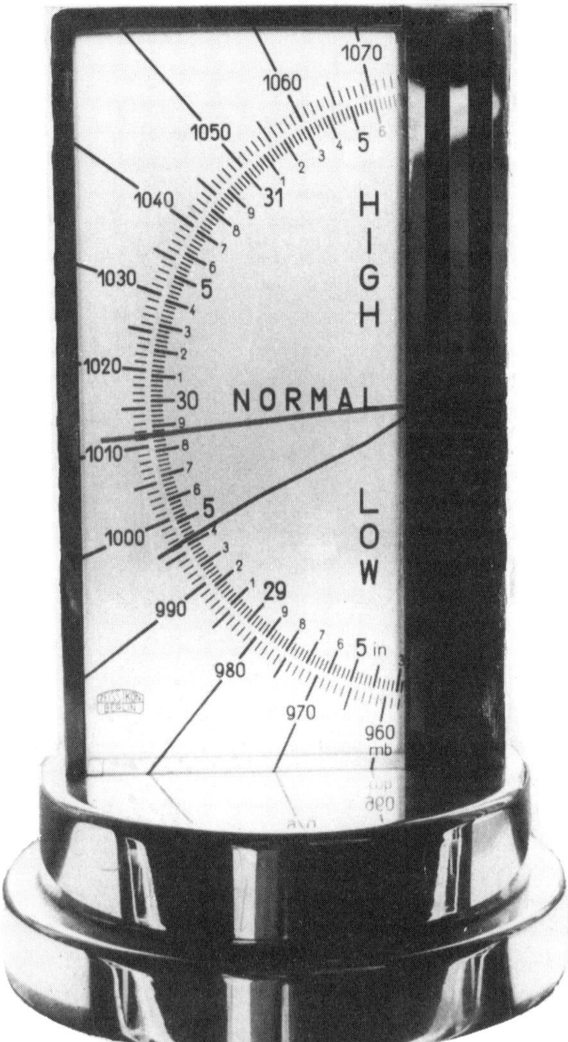
1070
1060
1050
1040
1030
1020
1010
1000
990
980
970
960
mb
31
30
29
HIGH
NORMAL
LOW
5 in

264

STORMY
RAIN
CHANGE
FAIR
VERY DRY

265

DRY
FAIR
30

266

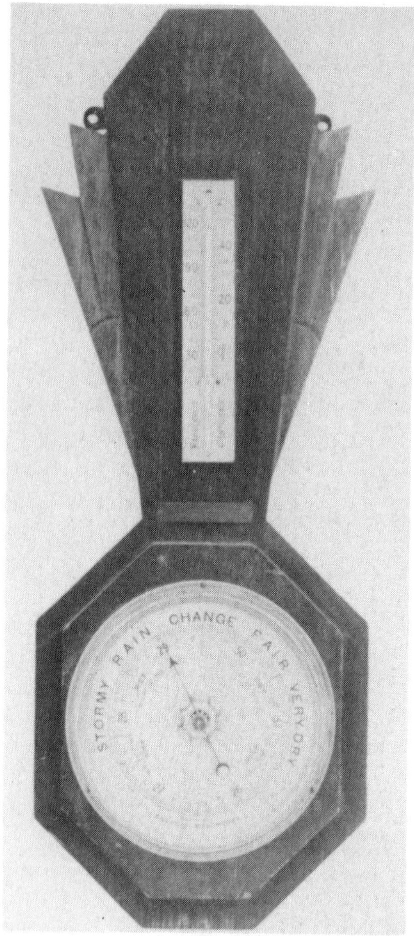

267

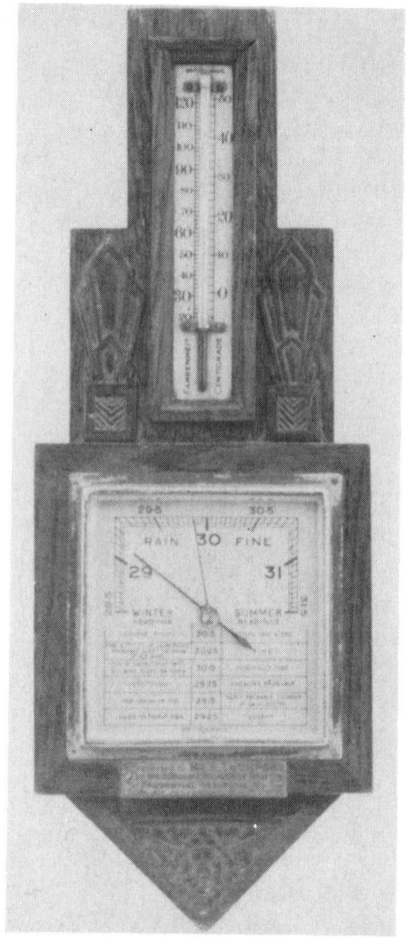

268

269

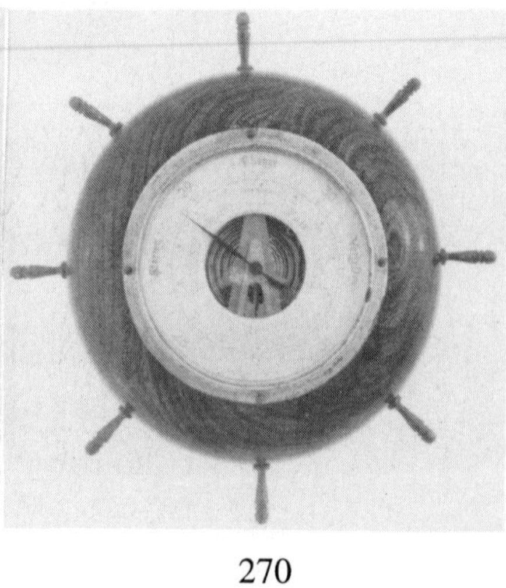

270

271

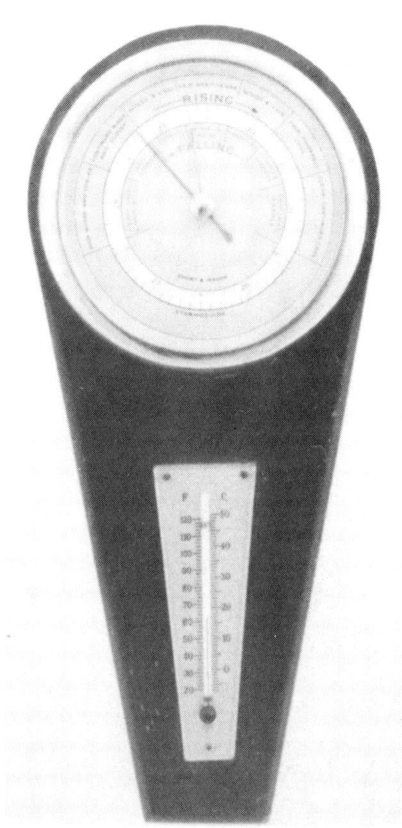

272

273

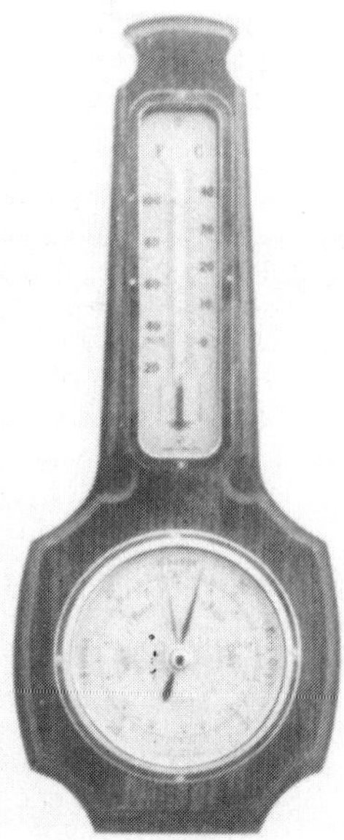

274

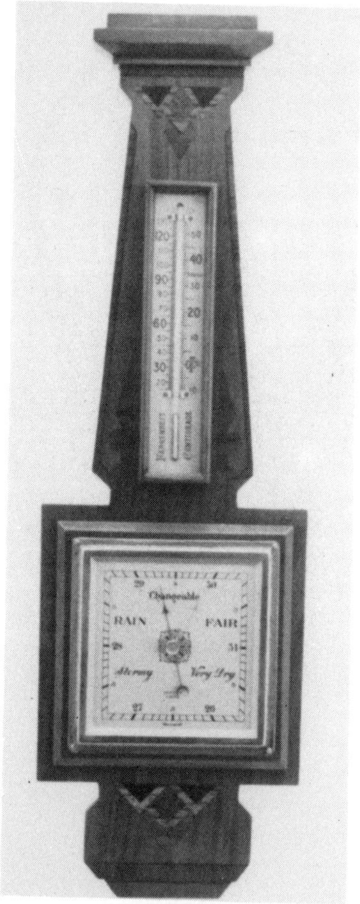

275

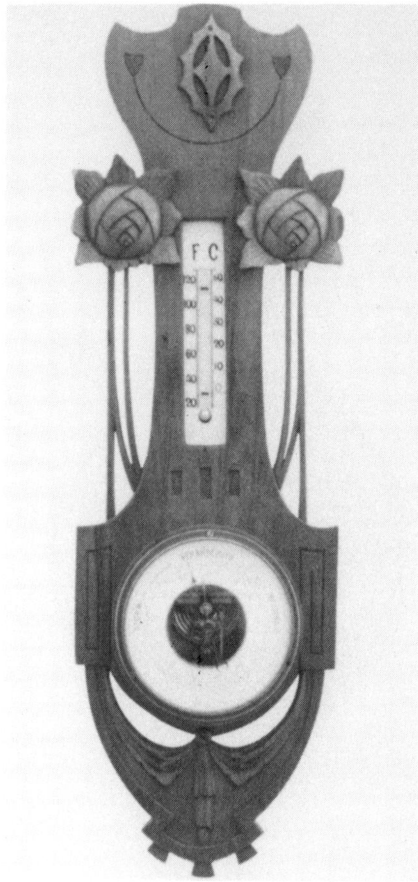

276

277

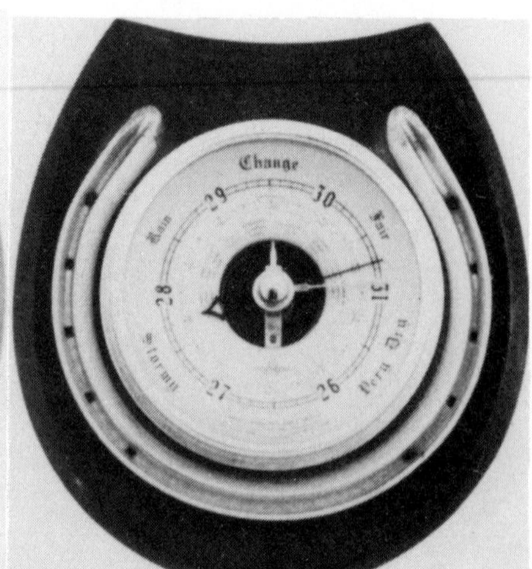

278

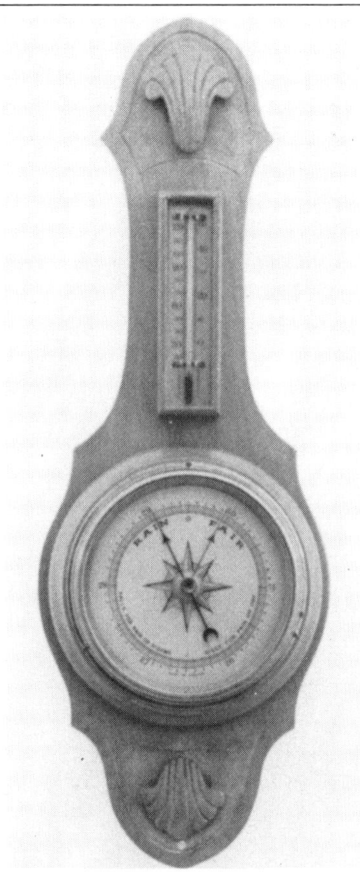

279

280

281

282

283

284

CHANGE
RAIN
FAIR
Much Rain
Set Fair
STORMY
VERYDRY

285

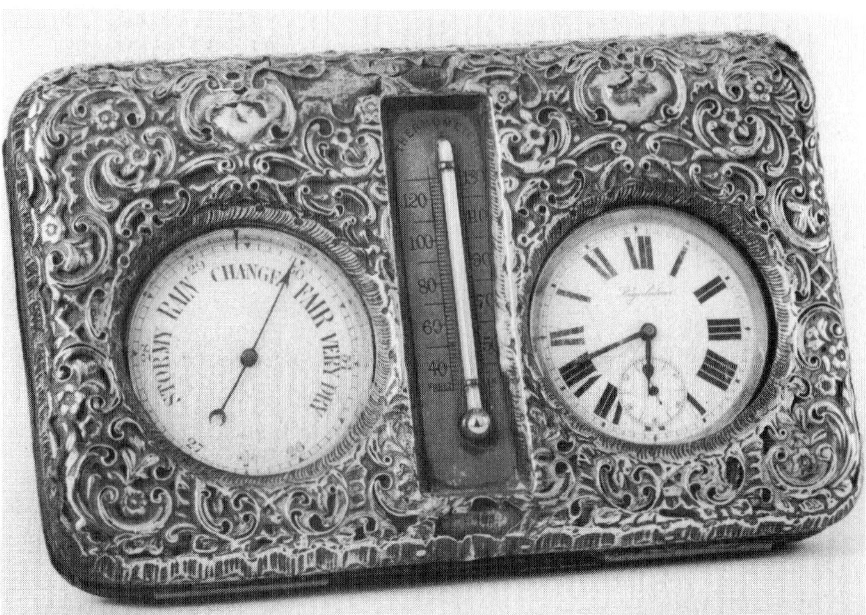
STORMY RAIN CHANGE FAIR VERY DRY

286

STORMY
MUCH RAIN
RAIN
CHANGE
FAIR
SET FAIR
VERY DRY

287

RAIN
CHANGE
MOTTERSHEAD & Co

288

289

290

291

292

293

294

295

296

297

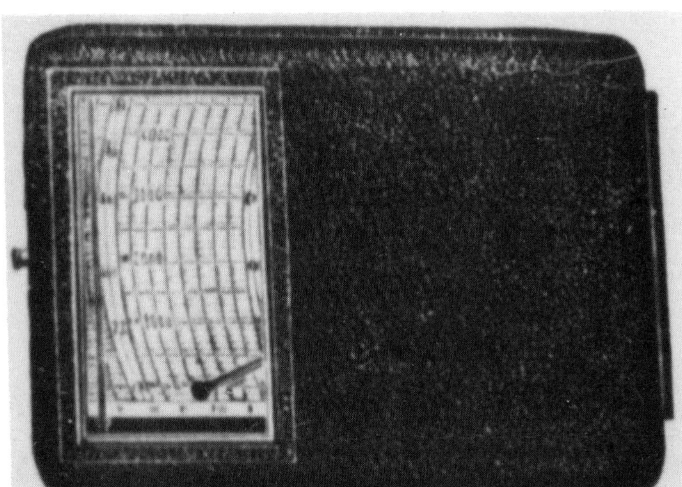

298

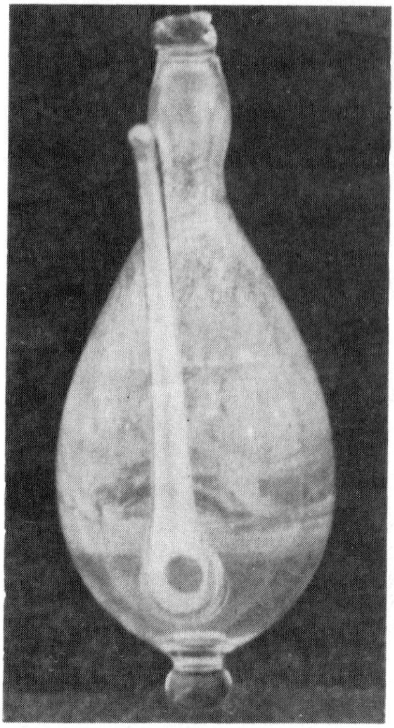
299

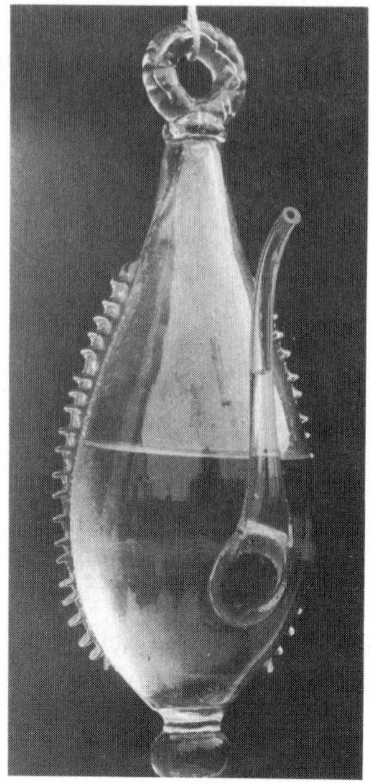
300

301

302

303

304

305

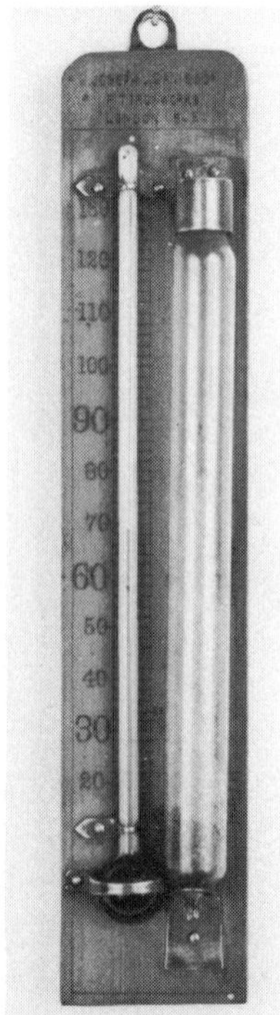

306

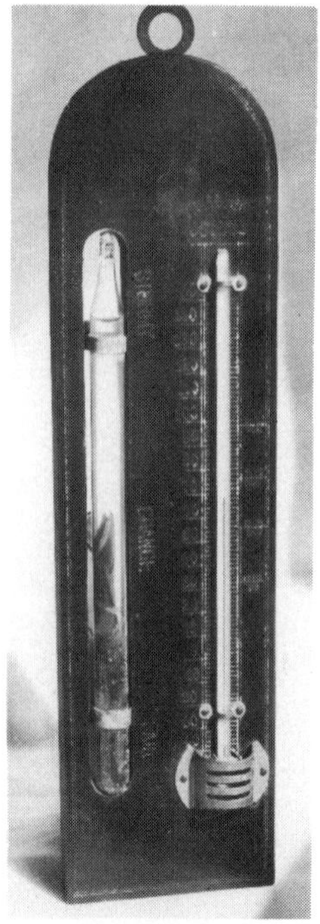

307

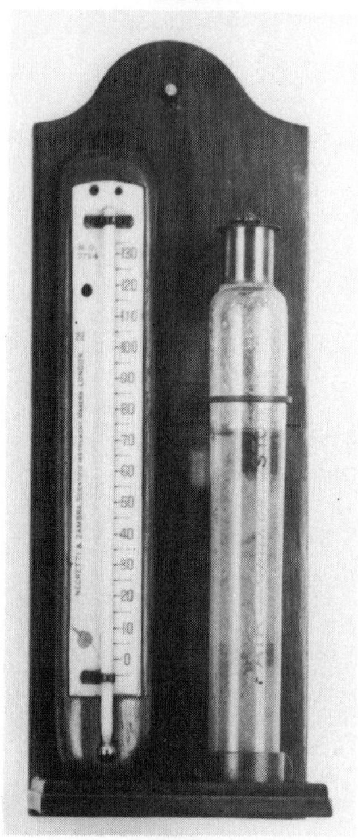

308

309

310

311

312

313

BRIEF NOTES ON THE ILLUSTRATIONS

The Museum Record Number is given in brackets at the end of the note.

1 Open cistern barometer in walnut case by Henry Wynne, c.1680. He was the first to make and sell barometers, in any number, to the general public from around 1675. (10)
2 Open cistern barometer in walnut case. Unsigned, c.1695. Many clockmakers diversified into barometer-making during this period and their influence on the case design can be clearly seen. (11)
3 Reproduction walnut pillar portable barometer in the style of Daniel Quare, London, c. 1700. He was one of the outstanding clock and barometer makers. (12)
4 Reproduction ebonised wood free-standing barometer in the style of Daniel Quare by Adams, London, c.1925. (1)
5 Walnut open cistern country barometer. Unsigned, c.1710. (13)
6 Mahogany portable cistern barometer by Edward Scarlett, London, c.1740. He also sold angle barometers and was 'Optician to His Majesty King George the Second.' (15)
7 Walnut cistern barometer with Royal Society scale thermometer. Unsigned, but probably by Stephen Davenport, London, c.1720. (14)
8 Detail of the register plates in 7 showing early eighteenth-century engraving.
9 Mahogany country barometer by James Verrier, North Curry, c.1750. The case design shows that he was basically a clockmaker and he developed a type of clock movement powered by barometric or temperature change. (16)
10 Detail of the register plates in 9 showing country-style engraving.
11 Mahogany bulb cistern barometer by Thomas Benbow, c.1750. (17)
12 Red walnut cistern tube barometer by James Ayscough, London, c.1755. A noted optician who occupied 'The Original Shop for superfine Crown-Glass Spectacles.' (18)
13 Mahogany cistern tube barometer by George Burton, London, c.1765. His instruments, including two barometers, were taken on Captain James Cook's second voyage of discovery in 1772. (22)
14 Mahogany and walnut open cistern tube barometer by John Bird, London, c.1760. A renowned instrument-maker noted for his accurate scales. The barometer tube is widened to a bulb at the top to dilute the effect of any air or vapour that might, in time, percolate into the vacuum. (19)
15. Detail of the register plates and hood in 14 typical of a clockmaker's design.
16 Portable cistern barometer in padouk wood by John Rittson, Tekrill, c.1760. A clockmaker who engraved his barometer register plates with the words 'The Great Barometer'. (20)
17 Detail of the register plates in 16 showing the fine engraving.
18 Mahogany triple weather-glass by Benjamin Martin, London, c.1760. The instument has a barometer, thermometer and hygrometer. Martin was an enterprising maker who held courses and public demonstrations; he also wrote over thirty popular scientific works. (21)

19 Detail of the register plates in 18 showing the barometer, thermometer and hygrometer scales.
20 Mahogany portable cistern barometer. Unsigned, c.1765. The most common design of barometer of this period. (23)
21 Mahogany barometer with cistern tube fitted with an ivory float and whalebone hygrometer by Dollond, London, c.1770. John Dollond was a Huguenot silk weaver and his son Peter became a brilliant instrument-maker, being appointed optician to King George III and the Duke of York. (24)
22 Mahogany barometer with oat beard hygrometer by Henry Pyefinch, London, c.1770. This type of barometer was advertised for sale by Pyefinch in 1770 for £3.3s. (25)
23 Mahogany barometer in protective case by Troughton, London, c.1770. The Troughton family were important instrument-makers, the most famous being Edward who never married; he was a man of frugal habits and spent most of his life in his back parlour wearing snuff-stained clothes and a wig, with an ear trumpet in his hand. (26)
24 Detail of the register plates in 23 showing the vernier adjustment mechanism.
25 Mahogany barometer with brass inlay by Edward Nairne, London, c.1770. He published various books on his navigational, astronomical and pneumatic instruments and was elected a Fellow of the Royal Society in 1776. (27)
26 Mahogany cistern barometer by Nairne & Blunt, London, c.1790. A partnership between Edward Nairne and Thomas Blunt. Blunt's barometers featured egg-shaped cistern covers and he became mathematical instrument-maker to George III. (28)
27 Mahogany barometer by Dollond, London, c.1780. An important family of instrument-makers; Peter was appointed optician to King George III and Duke of York. (29)
28 Mahogany barometer with a perpetual almanac by Francis Watkins, London, c.1780. The business was started in 1747 and was taken over by Elliott Bros in 1857; it is now part of the GEC-Elliott Automation Group. (31)
29 Detail of the perpetual almanac shown in 28.
30 Mahogany cistern barometer by George Adams, London, c.1780. The family were important instrument-makers and George succeeded his father, also named George, as instrument-maker to George III and later became optician to the Prince of Wales. (30)
31 Mahogany bulb cistern barometer by Balthazar Knie, Edinburgh, c.1790. The most well-known and prolific maker of barometers in Scotland. Born in Germany and spent several years as an itinerant barometer maker giving exhibitions of glassblowing before settling in Edinburgh. (32)
32 Mahogany barometer with round top and urn-shaped cistern cover by Smith, Bath, c.1790. (33)
33 Oak bulb cistern barometer with paper plates by Dominick Manticha, London, c.1780. One of the many Italians who migrated to England around 1780. (34)
34 Detail of the paper plates shown in 33.
35 Mahogany bulb cistern barometer with paper plates by Mathew Woller, Edgbaston Street, Birmingham, c.1800. The pagoda-type pediment suggests the Chinese Chippendale influence of 1765-70 and the pillared hood is a reminder of earlier clock-case designs. (35)
36 Mahogany cistern tube barometer with urn-shaped cover by Beilby, Bristol, c.1805. (37)

37 Mahogany barometer with square cistern cover by Edward Butler, Tutbury, c.1800. (38)

38 Mahogany barometer with bulb cistern and Tunbridge ware marquetry decoration round the trunk and cistern cover by A. Bellamy, Wycombe, 1799. (36)

39 Detail of register plates on 38 showing unusual wording for the weather indications.

40 Mahogany barometer with hygrometer and marquetry flower patterns by Baptista Ronchetti, Manchester, c.1800. A producer of high-quality instruments who arrived in Manchester in 1785 from Tavernerio, near Lake Como, Italy. (39)

41 Mahogany bulb cistern barometer with herring-bone veneer and chequered inlay by William Hardy, 171 Holborn, London, c.1810. (40)

42 Mahogany barometer with egg-shaped cistern cover by Thomas Blunt, London, c. 1805. He became mathematical instrument-maker to King George III. (42)

43 Mahogany barometer with thermometer and hygrometer by John Russell, Falkirk, c.1800. Oval cistern covers were popular with Scottish makers. Appointed watchmaker to the Prince Regent in 1811. (41)

44 Detail of the register plates in 43 showing the quality of the engraving.

45 Mahogany cistern barometer by Jones, 241 Oxford Street, London, c.1810. (43)

46 Mahogany bulb cistern barometer with hinged door and key to adjust the vernier by Joseph Pastorelli, 47 Alberon Street, Liverpool, c.1810. Two families named Pastorelli migrated from Italy to England around 1800; one family settled in London and the other in Liverpool. (44)

47 Mahogany bulb cistern barometer with hinged doors to silvered brass register plates and cistern. Sickle-shaped manual pointer by C. Zappa, Sheffield, c.1815. (45)

48 Satinwood barometer with hinged door to the concertina cistern with portable screw. The register plates are of ivory by Gardners, Glasgow, c.1822. A typical Scottish barometer but some have pointed tops rather than round. (46)

49 An inexpensive ordinary siphon tube barometer probably made for school or instructional use. Unsigned, c.1820. (47)

50 Mahogany bulb cistern barometer with thermometer having Fahrenheit and Reaumur scales by H. Hughes, 120 Fenchurch Street, London, c.1835. He established a Navigation Warehouse under the trade sign of 'Little Midshipman' which became a well-known rendezvous for seamen. (48)

51 Mahogany bulb cistern barometer in a square and moulded case by J. Pozzi, Oswestry, c.1825. An Italian maker whose brother Peter taught Italian in one lesson, for a guinea, in Watkin's, Oswestry. (49)

52 Mahogany bulb cistern barometer by Thomas Rubergall, 24 Coventry Street, London, c.1825. This barometer was at one time in the ownership of William Wordsworth. Rubergall was instrument-maker to the Duke of Clarence. (50)

53 Mahogany bulb cistern barometer with paper plates by Adam Routledge, Carlisle, c.1830. (51)

54 Mahogany barometer with bulb cistern cover decorated with marquetry. Paper plates signed by Charles Howorth, Halifax, c.1830. (52)

55 Detail of paper plates in 54 showing summer and winter weather indications.

56 Mahogany barometer with bulb cistern cover decorated with marquetry and inlays by George Purcheon, Glass House, Leeds, c.1830. (53)

57 Detail of paper plates in 56 shown to be almost identical to those in 55.
58 Mahogany bulb cistern barometer with painted register plates and marquetry decoration by George Purcheon, Glass House, Leeds, c.1830. (54)
59 Detail of metal plates in 58 painted and decorated with flower sprays coloured blue, brown, green, pink and red.
60 Mahogany bulb cistern barometer with herring-bone pattern veneer by Caesar Tagliabue, 23 Hatton Garden, London, c.1830. An Italian immigrant who became an important maker. The firm still trades today under the name of Casella London Ltd. (55)
61 Mahogany bulb cistern barometer with chequered stringing by Negretty & Co, London, c.1830. This was by far the most popular style of stick barometer of the period. (56)
62 Mahogany bulb cistern barometer with ebony and boxwood stringing which borders an intricate chequered inlay in the style of Tunbridge ware by Charles Aiano, 91 Northgate, Canterbury, c.1830. (57)
63 Heavily carved mahogany cistern barometer by Watkins & Hill, 5 Charing Cross, London, c.1840. The business was taken over by Elliott Bros in 1857 and is now part of the GEC-Elliott Automation Group. (58)
64 Rosewood barometer with ivory register plate by Negretti & Zambra, London, c.1850, but the retailer's name, W. Lund, 23-24 Fleet Street, London, is engraved on the register plates. (60)
65 Rosewood cistern barometer by W. & T.C. Heath, Devonport, c.1850. (61)
66 Mahogany cistern tube barometer with ivory plates, bow-fronted case and urn shaped cistern cover by William Holmes, London, c.1840. (59)
67 Detail of 66 showing the ivory register plates which, over a period, replaced silvered brass plates and scales.
68 Rosewood cistern barometer with two sets of weather indications and two verniers by R. Webb, Taunton, c.1860. (62)
69 Oak cistern barometer with porcelain register plates and shaped cistern cover by Elliott Bros, 449 Strand, London, c.1860. Elliott Bros is now part of the GEC-Elliott Automation Group. (63)
70 Oak agricultural or cottage barometer by L. Casella, London, c.1865. An important Italian instrument-maker. The firm still trades today under the name of Casella London Ltd. (64)
71 Oak cottage barometer by William Langford, 52 Broad Quay, Bristol, c.1865. (65)
72 Carved mahogany barometer by William Charles Cox, Optician, Devonport, c.1860. Cox also had a shop in Plymouth and supplied instruments to the Naval College. (66)
73 Miner's barometer in oak by J. Davis & Son, Derby & London, c.1875. The register plates are calibrated from 26 to 33 inches to allow the barometer to be used at 2,000 feet below sea level. (68)
74 Farmer's oak barometer with a wet and dry bulb hygrometer by Negretti & Zambra, London, c.1865. Two Italians who were prolific and leading makers of barometers and held appointments as instrument-makers to Queen Victoria. Still trading today as Negretti Automation. (67)
75 Detail of the porcelain register plates in 74 showing the wet and dry bulb hygrometer.
76 Oak model cistern barometer by James Hanny, Shrewsbury, c.1875. The most popular type of cheap barometer which sold for 13s 6d in 1875. (69)

77 Oak model cistern barometer by J. H. Steward, 54 Cornhill, London, c.1875. (70)
78 Oak cistern barometer by J. H. Steward, 54 Cornhill, London, c.1875. (71)
79 Carved oak cistern barometer by B. Boese, Kidderminster, c.1875. (72)
80 Mahogany double-angle barometer with bulb cistern by William Robb, Montrose, c.1800. A rare type of barometer which appears to have only been made by Robb and two other Scotsmen, Balthazar Knie and Charles Molliner. (74)
81 Mahogany angle barometer by John Whitehurst, Derby, c.1770. An important maker who moved to London in 1775 to become 'Stamper of Money Weights' at the Mint. (73)
82 Mahogany angle barometer with bulb cistern within a hinged door by Baptista Pedralio, c.1810. (75)
83 Mahogany double-tube angle barometer with paper register plates by Samuel Lainton, Halifax, c.1830. He also made single-tube angle barometers but all had paper plates. (76)
84 Mahogany marine barometer by Thomas Jones, 62 Charing Cross, London, c.1820. The barometer can be hung on a wall or mounted on a gimbals. (77)
85 Detail of the register plates in 84 showing the engraving and thermometer mounted on the inside of the door.
86 Mahogany marine barometer in cylindrical case decorated with spiral reeding and gadrooning. Ivory register plates signed C. A. Canti, 30 High Holborn, London, c.1860. (78)
87 Mahogany marine barometer with ivory register plates by J. Hicks, London, c.1870. An important maker and a staunch Catholic who presented meteorological instruments to the Vatican. (79)
88 Carved walnut marine barometer with a sympiesometer by John Campbell, 7 South Castle Street, Liverpool, c.1870. (81)
89 Carved mahogany marine barometer by C. G. Brander & Son, 82 Minories, London, c.1870. (80)
90 Detail of the ivory register plates in 89 showing the double weather indications and two verniers.
91 Kew pattern marine barometer in brass cylindrical case by Patrick Adie, 1 Broadway, Westminster, London, c.1885. A specialist in marine barometers, he was responsible for no less that 20 patents. (82)
92 Fitzroy marine or gun marine barometer in bronzed metal case. Unsigned, c.1900. (83)
93 Kew pattern marine barometer with a Gold Slide by Negretti & Zambra, London, c.1940. It is in its original carrying case of varnished deal with rope handles. (84)
94 Movable scale station barometer by J. Newman, 122 Regent Street, London, c.1840. Regarded as the most successful station barometer ever made and was installed in all the main meteorological observatories in the British Empire. (285)
95 Side view of the Newman barometer in 94.
96 Station iron cistern barometer in a brass case by J. F. Newman, 122 Regent Street, London, c.1840. A leading instrument-maker who made barometers for the Ross Antarctic Expedition. (85)
97 Detail of the silvered brass register plates in 96 showing the four corrections to improve accuracy. The weather indications can just be seen on the right-hand side of the case.
98 Standard barometer on Fortin's principle by Chancellor & Son, Dublin, c.1830. It has a glass cistern so that the level of the mercury can be seen. (86)

99 School Laboratory siphon tube barometer. Unsigned, c.1900. (88)
100 Balloonist's barometer in a mahogany case with an ivory scale which records down to 5 inches of air pressure by Henry Negretti, 19 Leather Lane, Holborn, London, c.1845. Negretti was an acquaintance of James Glaisher, the well-known balloon pioneer, and this barometer was probably used for his early balloon experiments. (89)
101 Fishery or sea coast barometer in an oak case by Ross, London, c.1865. (87)
102 Detail of the porcelain register plates in 101 showing the weather indications known as Admiral Fitzroy's scale words.
103 Great double barometer in a carved mahogany case. Unsigned, c.1760. (90)
104 Detail of the top section of the case showing the boxwood scale headed 'The Great Barometer' and ornamented with wheat-ear border and symmetrical designs.
105 Double barometer and thermometer in a mahogany case by T. Thomson, Edinburgh, c.1805. (91)
106 Double or contra-barometer and thermometer of the Art Nouveau period by Otto Baumbach, Manchester, c.1905. He was glassblower by appointment to the University of Manchester. (92)
107 Double or multiple-tube barometer and thermometer in a mahogany case with paper plates by Baptista Ronchetti & Co, Manchester, c.1785. An important Italian maker who arrived in Manchester from Tavernerio, Italy in 1785. (93)
108 Double or multiple-tube barometer and thermometer in mahogany case by Rabalio, c. 1795. The thermometer scales and weather indications are stamped on a boxwood base. (94)
109 Sympiesometer in mahogany case by Alexander Adie, Edinburgh, c.1825. This instrument was invented by Adie in 1818 and is numbered 537; it uses hydrogen and almond oil. Adie was optician to William IV and Queen Victoria. (95)
110 Improved marine sympiesometer in a carved mahogany case by James Hughes, Ratcliff, London, c.1870. (96)
111 Admiral Fitzroy's barometer in a carved oak case with thermometer and storm glass, c.1870. Fitzroy was a meteorologist and this type of barometer was named after him because he suggested its construction and his 'Remarks' on weather interpretation were used on the scale. (97)
112 Admiral Fitzroy's barometer in a carved oak case with thermometer and storm glass, c.1875 (98)
113 Admiral Fitzroy's barometer with storm glass in an oak case, c.1875. (99)
114 Admiral Fitzroy's barometer with thermometer in a mahogany case, c.1880. (100)
115 Admiral Fitzroy's barometer with storm glass and clock in an oak case by John G. Murdock & Co. Ltd of London and Melbourne, c.1885. (101)
116 Admiral Fitzroy's barometer with an iron tap below the tube bulb to make it portable, c.1890. (2)
117 Admiral Fitzroy's barometer and thermometer in a mahogany case with paper plates signed M. E. Solomons, Optician, Nassau Street, Dublin, c.1895. (102)
118 Royal Polytechnic barometer with two thermometers and two indicating hands in a carved oak case by Joseph Davis & Co London, c.1880. Sole manufacturers in England of this type of barometer. (103)
119 Wheel barometer in walnut case with silvered brass dial and recording dial by John

Hallifax, Barnsley, c.1730. A notable clock and barometer maker who produced distinctive instruments. (104)
120 Wheel barometer in decorated mahogany case with silvered brass dial and recording dial by J. Hallifax, Barnsley, c.1770. Probably made by George, the son of John, from the father's stock of parts after his death in 1750. George was twice Mayor of Doncaster. (105)
121 Mahogany 12 inch barometer with hygrometer attributed to Christopher Bettally, London, c.1785. One of the first Italians to arrive in London; he made distinctive barometers and also had a shop in Paris. (106)
122 Mahogany 8 inch barometer by James Gatty, London, c.1780, who made very high-quality instruments. He was one of the early Italian immigrants who introduced this design of case which was the most popular for late eighteenth century wheel barometers. (107)
123 Mahogany 8 inch barometer with thermometer, hygrometer and spirit level by J. M. Ronketti, 180 Holborn, London, c.1785. Another Italian immigrant who was a contemporary of James Gatty. Both were important makers. (108)
124 Broken pediment mahogany barometer with fan and floral paterae inlay by Lione & Co., Holborn, London, c. 1800. Probably Dominick Lione who came to London from Italy around 1800. (109)
125 Mahogany 8 inch barometer with unusual decoration by P. Gally, London, c.1800. Probably Paul or Peter Gally who were Italians working in Clerkenwell, London from the beginning of the nineteenth century. (110)
126 Detail of the dial in 125 showing the attractive engraving.
127 Mahogany 8 inch barometer decorated with flower head and zebra stringing by F. Amadio & Son, 118 St John Street Road, London, c.1800. This Italian family were prolific makers of high-quality barometers who settled in London around 1800. (111)
128 Satinwood 10 inch barometer by Francis Anone, 26 High Holborn, London, c.1805. An Italian immigrant who made high-quality instruments. (112)
129 Mahogany 10 inch barometer by Charles Pitsalla, 221 High Holborn, London, c. 1805. Another Italian immigrant who made barometers with pleasing lines. (113)
130 Mahogany 10 inch barometer with clock by Hudson, Greenwich, c.1800. Instrument-maker to the Admiralty, the Royal Naval College and the Royal Observatory. (114)
131 Mahogany barometer with 12 inch dial and clock by Mathew Woller, 51 Edgbaston Street, Birmingham, c.1805. The 12 inch dial was the largest made for domestic use. (115)
132 Satinwood barometer with 12 inch dial by Tagliabue & Torre, 294 Holborn, London, c.1805. Two important Italian makers. Satinwood was reserved for the best quality barometers. (116)
133 Mahogany barometer with thermometer on the dial by Lione & Somalvico, 125 Holborn Hill, London, c.1805. Two notable Italian makers. (117)
134 'Upside-down' mahogany wheel barometer with 10 inch dial by James Lione, 81 High Holborn, London, c.1810. A very rare type of barometer. (118)
135 Mahogany barometer with 8.25 inch dial by Chamberlain, Portsea, c.1810. Made in the style of, or as a companion to, a grandfather clock. (119)
136 Mahogany barometer with a thermometer on the 6 inch dial by Samuel Bregazzi, Queen Street, Derby, c.1810. A companion to a clock style barometer made by an Italian. (120)
137 Mahogany barometer with 8 inch dial and flower head and shell inlay by Joseph Cetti

& Co, 54 Red Lion Street, Holborn, London, c.1810. The Cettis were immigrant Italians who made a large number of quality barometers. (121)

138 Mahogany Sheraton shell barometer with 8 inch dial by Trombetta, c.1820. Probably Charles Trombetta of Norwich. A very large number of this style of barometer were made between 1815 and 1825. (122)

139 Mahogany Sheraton shell barometer with 8 inch dial and four shell inlays by Mazzuchi & Co, Gloucester, c.1820. (123)

140 Mahogany 8 inch barometer with broken pediment and urn motif marquetry by Lione & Somalvico, Holborn, London, c.1815. (124)

141 Mahogany 8 inch barometer with round top and urn motif marquetry by Lione & Somalvico, Holborn, London, c.1815. These two makers alone appear to have used the urn motif. (125)

142 Mahogany 8 inch barometer with flower petal decoration by G. Testi & Co., Chester, c.1820. (126)

143 Satinwood 8 inch barometer with broken pediment and ebony stringing by M. Salla, Preston, c.1815. (127)

144 Mahogany 8 inch barometer with flower head inlay by Caesar Tagliabue, 26 Holborn, London, c.1810. A prolific Italian barometer maker who arrived in London in the 1790s. (128)

145 Round-top 8 inch mahogany barometer by Peter Manticha. Holborn, London, c.1815. (130)

146 Mahogany 10 inch barometer with a stork depicted on the dial engraved by Charles Somalvico, 11 Brooke Street, Holborn, London, c.1815. (129)

147 Detail of the dial in 146 showing the bright-cut engraving and central cartouche over a stork.

148 Round-top 8 inch mahogany barometer by M. Bamforth, Leeds, c. 1820. (131)

149 Mahogany 8 inch barometer with cogged spindle adjustment to the set hand by a brass key fitted below the case by Francis Molton, Norwich, c.1815. (132)

150 Mahogany 8 inch barometer with shell and flower head inlay by Francis Molton, Norwich, c.1820. (133)

151 Mahogany 8 inch barometer in shield-shaped case by Angelo Alberti, Sheffield, c.1825. (134)

152 Satinwood 8 inch barometer with flower head inlay by Tognoni & Co., Shepton Mallet, c.1825. (135)

153 Mahogany 8 inch barometer with scroll pediment by Francis Amadio, 118 St John St Road, London, c.1830. The Amadio family were Italians and prolific makers of high-quality barometers. (136)

154 Mahogany 6 inch barometer by Adie & Son, Optician, Edinburgh, c.1840. Barometers with dials of this diameter are rare. Father and son held warrants as optician to William IV and Queen Victoria. (137)

155 Mahogany 4.5 inch barometer by F. Amadio & Son, 118 St John St Road, London, c.1835. The retailer's name, J. Mangiacavalli, 22 Charles Street, Hatton Garden, is engraved on the spirit level plate. Barometers with dials of this diameter are very rare. (138)

156 Rosewood 12 inch barometer with clock by S. J. M. French, Royal Exchange, London,

c.1830. He exported instruments to Spain and America. (139)

157 Mahogany 12 inch barometer with diurnal register by Robert Carr Woods, 47 Hatton Garden, London, c.1830. He was a founder member of the Meteorological Society. (140)

158 Detail of the diurnal register in 157 which is used to record daily readings at a particular time. It is adjusted by a cog mechanism through the ivory key on the right.

159 Mahogany 8 inch barometer in squared frame edged with crossbanding and stringing by A. Gilardoni, Bristol, c.1840. (141)

160 Mahogany 12 inch barometer in squared frame with crossbanding and stringing by C. Maspolli, Manchester, c.1840. (142)

161 Rosewood 8 inch barometer with mother-of-pearl inlay by Paul Carughi, London, c.1840. (143)

162 Rosewood 10 inch barometer with brass stringing by Edward Barrett, Blandford, c.1840. (144)

163 Mahogany 10 inch barometer with squared frame and brass pillars flanking the thermometer by Gardner & Dowling, Belfast, c.1840. (145)

164 Mahogany 10 inch barometer with squared frame and mahogany pillar each side of the thermometer by McDowall, Edinburgh, c.1845. (146)

165 Mahogany 10 inch barometer with round top and moulded base by Dollond, London, c.1845. The Dollonds were an important family of instrument-makers. This barometer was originally constructed to hang with a moulded top and a round base. (147)

166 Mahogany 8 inch barometer with a mirror by John Schalfino, Taunton, c.1845. This was the most popular type of wheel barometer in the 1840s. There is an engraved masonic sign on the dial; these appear on most of his barometers. (148)

167 Mahogany 10 inch barometer with mirror and thermometer on the dial by J. Spelzini, 11 Brooks Market, Holborn, London, c.1840. (150)

168 Oak 10 inch barometer by Mansford, London, c.1845. (149)

169 Detail of the dial in 168 showing the engraving style of the mid-nineteenth century and masonic sign above the central cartouche. The ivory plate above the dial is inscribed 'Made from the wreck of the Royal George. Sunk, August 29th 1782'. It was raised in 1839.

170 Mahogany 8 inch barometer with rectangular mirror and dial engraved with a windmill scene by J. Laffrancho, Ludlow, c.1840. He alone appears to have made this style of barometer. (151)

171 Oak stick and 10 inch wheel barometer using a single tube by Nathaniel Whitehouse, Leicester Square, London, c.1840. (152)

172 Mahogany 8 inch barometer in a squared case by Watkins & Hill, 5 Charing Cross, London, c.1848. A cork float connected to a small toothed brass bar and cogwheel is used instead of a pulley and glass weights. (153)

173 Mahogany pillar barometer with a siphon tube contained within the fluted column by Brace Girdle, London, c.1850. (154)

174 Detail of the open dial in 173 with the centre supported by a gilt metal male figure.

175 Mahogany 8 inch four-dial barometer with onion-top design. Unsigned, c.1850. (286)

176 Rosewood 8 inch onion-top five-dial barometer by Peter Chick, Worthing, c.1855. This was the first purchase of the collection and cost £6 unrestored in 1971. (155)

177 Rosewood 8 inch barometer with gold paint to simulate brass stringing and leaf

decoration by P. Bragonzi, Hereford, c.1855. (156)
178 Rosewood 8 inch barometer with pearl, tortoiseshell and brass inlay decoration by Gardner & Co., Glasgow, c.1858. Noted barometer makers in Scotland. (157)
179 Oak 10 inch barometer with carved scrolls by John Marratt, King William Street, London Bridge, c.1858. Exhibited at the Great Exhibition at the Crystal Palace in 1851. (158)
180 Rosewood 10 inch barometer with mother-of-pearl inlay by Charles Hodson, Worcester, c.1860. Hodson was also a clockmaker. (159)
181 Walnut 10 inch barometer in a carved case by Moritz Pillischer, 88 New Bond Street, London, c.1857. Won a medal at the Great Exhibition at the Crystal Palace in 1851. (160)
182 Walnut 8 inch barometer with round moulded edges by Negretti & Zambra, London, c.1870. The dial is engraved 'Instrument makers to Her Majesty'. (161)
183 Rosewood 10 inch barometer with pewter inlay by James Peters, Cambridge, c.1870. (162)
184 Oak 8 inch barometer with moulded edges and porcelain dials by Richard Howse, Marlborough, c.1875. (163)
185 Walnut 8 inch barometer with painted dial. Unsigned, c.1890. (164)
186 Reproduction mahogany 8 inch barometer. Unsigned, c.1900. (165)
187 Rosewood pendulum clock and 8 inch barometer with circular thermometer by Peter Bregazzi, Nottingham, 1843. This instrument was patented at the British Patent Office in 1843 and has the Royal Coat of Arms engraved on the dial. (167)
188 Bourdon metallic 7.5 inch barometer in an ebonised case made under E. Bourdon and Richard's Patent, Paris, c.1865. (168)
189 Bourdon metallic 4.5 inch barometer in a brass case, c. 1870. (253)
190 Vidie early aneroid barometer in a brass case with curved thermometer by E. J. Dent. Paris, c.1850. (231)
191 Early Vidie marine barometer with 4.5 inch dial and curved thermometer in an oak box by E..J. Dent, c.1850. (236)
192 Watch-sized barometer with 1.75 inch enamel dial in a brass case, c.1865. (171)
193 Brass pocket barometer with l.75 inch dial mounted with a compass, c.1865. Inscribed on the reverse side is 'For progress in Navigation HMS Britannia 1866'. (240)
194 Vidie aneroid barometer with printed 4.5 inch card dial in a brass case, c. 1860. (169)
195 Open dial barometer of 6 inch diameter in a brass case on a carved oak table stand by E. G. Wood, London, c.1865. (3)
196 Closed glazed card 4.5 inch dial barometer in a brass case on a carved and ebonised oak table stand, c.1865. (170)
197 Watch-sized barometer with 1.75 inch open dial with an altitude scale up to 10,000 feet in a velvet-lined leather case, c.1870. (173)
198 Pocket-sized barometer with 2.5 inch silvered dial which records heights up to 10,000 feet by Negretti & Zambra, London, c.1865. Prolific and leading makers of aneroid and other barometers. (4)
199 Watch-sized barometer with 2 inch enamel card dial which records heights up to 10,000 feet, c.1870. (172)
200 Traveller's barometer with a 2 inch silvered metal scale from 31 to 18 inches giving an altitude range of 15,000 feet housed in a silk-lined leather-covered case, c.1870. (174)

201 Watch-sized barometer with a 1 inch dial and altitude scale up to 10,000 feet in a shut-up hunter case, c.1870. (175)
202 Surveying barometer with 2.5 inch silvered dial and movable magnifier, 7,000 feet altitude scale with carrying case by James Hicks, London, c.1870. One of the leading barometer makers in the United Kingdom. (176)
203 Open dial 6 inch barometer with two curved thermometers housed in a brass case by Salom & Co., London and Edinburgh, c.1870. (177)
204 Brass 4.5 inch barometer with Admiral Fitzroy's words on weather on the dial by Manning & Co., Worcester, c.1870. (178)
205 Aneroid barometer in a walnut case with carved scrolls by Marratt & Ellis, 63 King William Street, London Bridge, c.1875. (247)
206 Oak aneroid barometer carved in the Gothic manner. Enamel card 5 inch dial, c.1875. (179)
207 Aneroid barometer with 8 inch dial in oak case with applied scroll carving, c.1880. (249)
208 Stem wind 1.5 inch barometer with 10,000 feet altitude scale in a silk-lined leather-covered case, c.1885. (185)
209 Carved oak barometer with a 5 inch dial. The case has a castellated top; intended for the wall of a ship's cabin, c.1875. (180)
210 Painted wall barometer with an open enamel card dial, c.1875. (5)
211 Patent rapid indicating aneroid barometer with 4.75 inch white glass dial, c.1880. (255)
212 Silver watch-sized barometer with 1.25 inch dial by J. H. Steward, 456 West Strand, London, c.1880. (238)
213 Fisherman's aneroid barometer with 5 inch enamel dial issued by the Royal National Lifeboat Institution and made by Negretti & Zambra, London, c.1890. (233)
214 Fisherman's aneroid barometer with 4.75 inch ceramic dial issued by the Royal National Lifeboat Institution and made by Dollond, London, c.1890. (234)
215 Marine aneroid barometer in a plain oak case and with a 5 inch porcelain dial. Made specially for the Shipwrecked Fishermen & Mariners Royal Benevolent Society by Dollond, London, c.1880. (181)
216 The Universal Barometer with 5 inch painted metal dial. A robust instrument with an iron frame and stout brass bezel for outdoor use, c.1880. (182)
217 Atmos barometer in a mahogany case 7 inches high and with curved thermometer, c.1880. This is an aneroid barometer. (183)
218 Pocket barometer with an enamel dial housed in a standing leather case, c. 1885. (186)
219 Watch-sized barometer with 1.75 inch silvered metal dial and height scale up to 8,000 feet by Chadburns Ltd, Liverpool, c.1880. (184)
220 Pocket barometer with 1.75 inch dial in gilded case. Thermometer with indicating pointer mounted on the dial by Watson Bros, 31 Cockspur Street, London, c.1894. (239)
221 Tourist set of barometer, thermometer and compass in a hinged leather case made in Geneva, c.1885. (187)
222 Pocket barometer with thermometer and compass in a double-hinged leather case by G. Lufft, Stuttgart, c.1885. See 225 for the reverse side. (188)
223 Pocket barometer with compass in a gilt metal case enclosed in a double-hinged

leather case by Eardley B. Norton, Manchester, c.1885. See 226 for the reverse side. (242)
224 Watch-sized barometer with 1.75 inch dial with thermometer and compass in a gilt metal case. Altitude scale to 20,000 feet at around 15 inches, c.1885. See 227 for the reverse side. (243)
225 Reverse side of barometer in 222 showing thermometer and compass.
226 Reverse side of barometer in 223 showing compass.
227 Reverse side of barometer in 224 showing compass.
228 Surveying or mining engineer's barometer with 4.5 inch dial and magnifier. The range is from 4,000 feet above to 2,000 feet below sea level by L. Casella, London, c.1890. Prolific makers of all types of instruments. (189)
229 Carved oak marine barometer with 6 inch porcelain dial, c.1890. (190)
230 Decorated ceramic cased barometer with gilded bezel. Hand-engraved 4.5 inch silvered dial by Negretti & Zambra, London, c.1890. (252)
231 Carved oak barometer with spiral thermometer. Aneroid 8 inch dial by Albert E. Dark, Barnstaple, c.1890. (254)
232 Carved oak aneroid barometer with 8 inch ceramic dial by Heath & Stoneman, Opticians, Plymouth, c.1890. (258)
233 Carved continental open dial aneroid barometer, c.1890. (250)
234 Banjo-shaped novelty barometer with painted brass front and open card 3 inch dial. Spanish weather indications replaced by English, c.1900. (7)
235 Art Nouveau carved barometer with 3.5 inch open porcelain dial, c.1905. (191)
236 Carved mahogany hall barometer with 4.5 inch open porcelain dial, c.1915. (192)
237 Profusely carved oak barometer with 5 inch enamel card dial, c.1890. (6)
238 Carved continental barometer with 4.25 inch open dial, c.1920. (260)
239 Pocket barometer in a gilt case with a silvered metal 2.5 inch dial and raised altitude scale up to 10,000 feet by John Wardale & Co., London, c.1918. (193)
240 Surveyor's 1.75 inch barometer with altitude scale up to 8,000 feet and down to 1,000 feet below sea level by J. H. Steward, London, c.1915. (194)
241 Travelling set with 1.5 inch barometer, thermometer and compass in double-opening morocco case, c.1915. (195)
242 Closed case of travelling set in 241.
243 Traveller's barometer with 1.75 inch dial in silver case. Watkins patent No. 374 by J. Hicks, London, c.1915. (237)
244 Brass weather forecaster by Negretti & Zambra, London, c.1915. (196)
245 Ivory pocket weather forecaster by Negretti & Zambra, c.1915. (244)
246 Reverse side of weather forecaster in 245 showing 26 different weather forecasts.
247 Weather watch or pocket weather foreteller with 1.75 inch dial by Negretti & Zambra, London, c.1920. (198)
248 Side and rear view of the weather watch in 247 showing the altitude scale and the 26 weather forecasts.
249 Bronze dome-shaped aneroid barometer with an overall diameter of 4.75 inches, c.1930. (200)
250 Oak marine barometer with 6 inch porcelain dial by John Barker & Co. Ltd, London, c.1925. (199)
251 Plain oak barometer with open card 3 inch dial, c.1930. (8)

252 Plywood and oak wall barometer, c.1920. (274)
253 Oak barometer with 6 inch dial by Budenberg Gauge Co., Broadheath, Altrincham, c.1935. (246)
254 Mahogany aneroid barometer inlaid with flower and leaf motifs by Page Keen & Page Ltd, Plymouth, c.1929. (259)
255 Mahogany aneroid barometer with 8 inch dial by F. Hilser, Trowbridge, c.1935. This barometer belonged to my father. (201)
256 Carved American walnut barometer with open dial, c.1920. This barometer was made on the Continent. (257)
257 Continental barometer with 3 inch open dial, carved in the shape of a sunflower, c.1930. (256)
258 Carved oak barometer with 3.5 inch open dial, c.1930. (262)
259 Aluminium and chromium desk barometer, c.1930. (278)
260 Barometer and rototherm thermometer in oak and plywood case by Shortland & Bowen, c.1930. (287)
261 Walnut stormoguide barometer with 11.25 inch dial by Short & Mason, London, c.1930. (282)
262 Stormoguide barometer in red painted bakelite case by Short & Mason, c.1930. (279)
263 Mahogany desk barometer by Short & Mason, London, c.1930. (283)
264 Chromium-plated table barometer with inch and millibar scales, 6.25 inches in height and made for the British market by Zeiss Ikon, Berlin, c.1935. (202)
265 Weathercraft aneroid barometer in perspex case, c.1930. (280)
266 Desk aneroid barometer, c.1950. (284)
267 Oak aneroid barometer with chromium bezel, c.1944. (272)
268 Presentation carved oak barometer, c.1946. (271)
269 Oak barometer with 4.5 inch square dial, c.1955. (261)
270 Elm barometer with brass spokes by Shortland Smiths Instruments, c.1950. (276)
271 Wall barometer with 3 inch open dial, c.1950. (275)
272 Stormoguide aneroid barometer and thermometer by Short & Mason, London, c.1950. (277)
273 Open dial barometer with mirror and brush set, c.1950. (270)
274 Elm barometer with 3.75 inch dial by Shortland Smiths Instruments, c.1955. (266)
275 Walnut aneroid barometer with marquetry decoration by Brufords, Exeter, c.1960. (268)
276 Carved Continental open 4.25 inch barometer, c.1960. (267)
277 Elm horseshoe barometer with 3.75 inch dial by Shortland Smiths Instruments, c.1960. (265)
278 Mahogany horseshoe barometer by Shortland & Bowen, c.1965. (245)
279 Carved barometer with 4.5 inch dial by 0. Comitti, London, c.1970. (269)
280 Meteorological Office 4.5 inch barometer Mark I with inch and millibar scales in a brass cylindrical case by Short & Mason, London, c.1945. (203)
281 Meteorological Office 4.5 inch barometer Mark II with millibar scale in a brass cylindrical case by Short & Mason, London, c.1966. (232)
282 Meteorological Office 4.5 inch barometer Mark II with millibar scale by Shortland Bowen Instruments, c.1970. (9)

283 Clock, barometer and thermometer in painted cast iron case with an overall height of 26 inches by J. J. Wainright & Co., Birmingham, c.1875. (204)
284 Clock, barometer and thermometer in cast iron case, c.1880. (248)
285 Clock, barometer and thermometer on an ebonised wood mounting by Negretti & Zambra, London, c.1875. (205)
286 Pocket aneroid barometer and pocket watch with thermometer in a silver standing case, c.1885. (206)
287 Carriage clock, barometer and compass in brass case. The compass is above the barometer, which, like the clock, has a porcelain dial, c.1905. The barometer dial has the wording 'By special appointment to H. M. the King.' (207)
288 Self-recording aneroid barometer with clock in lacquered case by Mottershead & Co., Manchester, c.1875. (208)
289 Barograph in oak case with drawer for storing the charts, c.1885. (209)
290 Barograph in walnut case, c.1885. (210)
291 Barograph in mahogany case with drawer for storing the charts, c.1890. (211)
292 Thermobarograph in oak case with chart drawer, c.1920. The instrument records air pressure and temperature. (212)
293 Barograph in oak case, c.1915. (213)
294 Barograph with Cyclo-Stormograph and weather forecasts in oak case by Short & Mason, London, c.1920. The vacuum chamber is housed under the base plate. (214)
295 Portable barograph in mahogany case by Short & Mason, London, c.1920. (215)
296 Micro-barograph in stove enamelled case by Short & Mason, London, c.1943. There is a higher bank of vacuum chambers to amplify the pen movement. (216)
297 Barograph with circular scale in an electro-plated nickel silver case, c.1925. (217)
298 Recording pocket barograph in morocco covered metal case 4.75 x 3.25 x 1.5 inches, c.l910. (229)
299 Old Dutch weather glass. Also known as Liege barometers and in Holland as the Donderglas. (218)
300 Old Dutch weather glass. The frills along the sides of the glass act as a kind of graduation. (219)
301 Water glass barometer, c.1860. (220)
302 Glass desk barometer by C. P. Goerz, Berlin, c.l925. (251)
303 Carved oak weather house, c.1880. (221)
304 Kendella weather house, c.1935. (226)
305 Weather house, c.1955. (227)
306 Storm glass and thermometer by Joseph Davis & Co., Fitzroy Works, London, c.1870. (222)
307 Storm glass and thermometer on an ebonised wood board, c.1880. (223)
308 Storm glass and thermometer on a mahogany window bracket. Storm glass by J. Blount Thomas & Co., Southampton, c.1875. (224)
309 Boxwood Fahrenheit thermometer by Negretti & Zambra, London, c.1920. (235)
310 Brass altimeter with 2.25 inch dial and leather strap by F. E. Becker & Co., Hatton Wall, London, c.1915. (241)
311 German recording altimeter in a birch case by E. Scholz, Hamburg, c.1915. (228)
312 Altimeter with scale up to 20,000 feet, c.1940. (230)
313 Hair hygrometer with 4 inch porcelain dial by Negretti & Zambra, London, c.1935. (225)

INDEX OF MAKERS' NAMES GIVING ILLUSTRATION NUMBERS